T0336149

The Unity of Science and Economics

Jing Chen

The Unity of Science and Economics

A New Foundation of Economic Theory

 Springer

Jing Chen
School of Business
University of Northern British Columbia
Prince George, BC
Canada

ISBN 978-1-4939-3464-5   ISBN 978-1-4939-3466-9   (eBook)
DOI 10.1007/978-1-4939-3466-9

Library of Congress Control Number: 2015954963

Springer New York Heidelberg Dordrecht London

© Springer Science+Business Media New York 2016
This work is subject to copyright. All rights are reserved by the Publisher, whether the whole or part
of the material is concerned, specifically the rights of translation, reprinting, reuse of illustrations,
recitation, broadcasting, reproduction on microfilms or in any other physical way, and transmission
or information storage and retrieval, electronic adaptation, computer software, or by similar or dissimilar
methodology now known or hereafter developed.
The use of general descriptive names, registered names, trademarks, service marks, etc. in this
publication does not imply, even in the absence of a specific statement, that such names are exempt from
the relevant protective laws and regulations and therefore free for general use.
The publisher, the authors and the editors are safe to assume that the advice and information in this
book are believed to be true and accurate at the date of publication. Neither the publisher nor the
authors or the editors give a warranty, express or implied, with respect to the material contained herein or
for any errors or omissions that may have been made.

Printed on acid-free paper

Springer Science+Business Media LLC New York is part of Springer Science+Business Media
(www.springer.com)

To Jamie Galbraith

Preface

Some ideas from established economic theories, such as perpetual growth or sustainable growth, are not consistent with basic scientific principles. Could better policies be developed to deal with today's social and economic problems if economic theories were more scientifically consistent? Economic and social policies have many dimensions. Should tax rate be higher or lower? Should interest rate be higher or lower? Should retirement age be later or earlier? Should the number of years of mandatory education be increased or decreased? Can the magnitude of business cycles and financial crises be reduced? Should government regulate business activities more or less? Does the creation of a Euro zone benefit or harm Europe? It seems that answering each question requires very specialized knowledge. In general, most people feel that social and economic problems are too complex to be understood from a simple theory.

To this, we may reflect on the evolution of our thoughts about celestial bodies. Before the development of modern astronomy, celestial bodies, which are very far away from us, were much more mysterious than the earthly matters, which are very close to us. Then, Nicholas Copernicus showed that when the sun, instead of the earth, was considered as the center of the universe, the trajectories of the planets looked simpler. Later Johannes Kepler discovered that the trajectories of planets were simple elliptic curves around the sun. The simplicity of these trajectories suggested to people that the movements of the planets may be governed by simple rules. Eventually, Newton showed that the movement of celestial bodies is indeed governed by the simple gravitational law. After that, we began to feel that celestial systems are much simpler than social systems. The level of complexity of a system often depends on how the system is described. Currently, the standard economic theory is dominated by general equilibrium theory. The idea of equilibrium has a long tradition in human society. In Bible, God created the world in six days. After that, God only intervened occasionally, such as flood in Noah's time, to restore equilibrium. However, since Darwin, scientists have abandoned this equilibrium theory about life. Today, researchers in science generally understand biological systems, which include human societies, as non-equilibrium systems. In this book, we will show that when human societies are described as non-equilibrium systems, economic activities become much simpler to understand.

Biological and social systems are indeed very complex. But beneath this complexity lies two common properties. First, all life systems need to obtain resources from the environment to compensate for the continuous dissipation to maintain life. Second, for any life form to be viable, its cost to obtain resources cannot exceed the value of the resources over its life cycle. Similarly, for a business to be viable in the long term, its average cost of operation cannot exceed its revenue. Costs include fixed cost and variable cost. The first property is a physical principle, and the second property is an economic principle. In short, all organisms and organizations need to satisfy a physical principle and an economic principle. From these two principles, we develop a mathematical theory of the relations among the main factors in economic activities, such as fixed cost, variable cost, duration of operation or life span of organisms, uncertainty, discount rate, and level of output. Due to their importance, these major factors in production naturally became the center of investigation in the early economic literature. However, because of the difficulty in forming a compact mathematical model about these factors, discussion about them became peripheral in the current economic literature. With the help of the analytical production theory, theoretical investigation in economics may refocus on important issues in economic activities. This theory greatly simplifies our descriptions of the structures and functions of human societies. It enables us to systematically analyze the return of biological and social entities with specific structures in specific environment. It enables us to perceive clearly about the long-term consequences of personal choices, economic policies, and social structures. In a recent book, *The End of Normal*, James Galbraith discussed many ideas related to this theory in great clarity.

Economic activities are based on human decisions. Any sound economic theory has to be established on a sound theory of mind. Currently, human mind is often described as "rational" or "irrational" in the economic literature. But there is no objective measure of being "rational" or "irrational." Human mind, as part of the human body, is evolved under the same economic principle that its average cost has to be less than its average value. More than one hundred years ago, Maxwell wondered, if the cost of information processing is less than the reduction of entropy from the information, the second law of thermodynamics will be violated. Because he felt that the second law is a very fundamental law, he concluded that the average cost of information processing must be higher than the average value of information, measured in terms of entropy. If this is true, how it is possible that the human mind, an information processing system, can generate a surplus? This is because the world is not entirely random. If some patterns in life are very common, they will become imprinted into human mind, becoming part of our instincts. So we do not have to reanalyze similar situations from the scratch. Instead, we respond automatically, which greatly reduces the cost of information processing.

An economic analysis suggests that the cost of information processing of our mind has to be lower than its value. How to measure the cost and value of information processing by humans and other living systems? Maxwell linked the cost and value of information to entropy. In 1948, Claude Shannon formally defined information mathematically as entropy. From the thermodynamic theory, entropy flow drives most directional movements, including movements in human societies. Hence, entropy provides a universal measure of value. However, one's subjective assessments often differ from the objective distributions and the prevailing opinions. To develop a good theory of human mind, we need to represent objective distributions, one's subjective assessments, and the prevailing opinions properly. Guided by intuition from statistical mechanics, we apply several functions generalized from the entropy function to measure the costs and values of information processing and decision making with different subjective assessments, objective distributions, and prevailing opinions. The resulting theory of mind may be called the entropy theory of mind. Since entropy provides a natural measure of value and cost for living systems, an entropy theory of mind is also an economic theory of mind. Recently, there have been many attempts to establish a behavioral theory of economics and finance. Instead of constructing a behavioral theory of economics directly, we develop an economic theory of behavior. Then, we integrate the value and cost of information processing into the overall picture in human decision making. The theory provides a quantitative link between our judgment and decision making, such as trading activities by investors. It offers simple and consistent descriptions of many patterns of asset market and investor behaviors that have puzzled the researchers. More generally, thinking and learning are guided by the consideration of value and cost of these processes. The entropy theory of mind, as an economic theory, provides a simple description about basic patterns of learning and human psychology.

The theory of mind is derived from the combination of economic and physical principles, just like the theory on the relation of major factors in economic and biological systems. Overall, the whole theory is an integration of economic and physical principles. Entropy provides a natural measure in both physics and economics. George Williams, an evolutionary biologist, once stated, "A biological explanation should invoke no factors other than the laws of physical science, natural selection, and the contingencies of history." The rate of return, an economic measure, simply provides a quantitative measure of natural selection. The contingencies of history are a consequence of fixed costs, which is a necessity from physical and economic principles. So our theory is consistent with George Williams' vision of a biological explanation.

Physical theories emphasize the relations among observable quantities.

> Today, there are not a few physicists who ... regard the task of physical theory as being merely a mathematical description (as economical as possible) of the empirical connections between observable quantities ... without the intervention of unobservable elements. (Schrodinger 1928, p. 58)

However, today's economic theory is mainly built on unobservable elements. Individuals are supposed to maximize "utility." Most mainstream economists believe that problems in economic activities are caused by "imperfect" competition. Yet patent laws and other legal measures are developed to promote monopoly over property rights, hence "imperfect" completion. These same economists believe that markets are the most "efficient" way to allocate resources. Nonetheless, laws and regulations are required because of "externalities" or "market failure." But government interventions often generate "government failure." Human beings are "rational" most of the time. However, stock market can turn very volatile because investors can become "emotional" or "irrational." Rarely, fundamental concepts in established economic theories are based on observable quantities.

This economic theory is built on observable quantities. We will not judge whether a person, a business, or a social system maximizes its "utility." We will not discuss whether competition is "perfect." Instead, we will only measure the return from a system with specific structures. If a system makes negative return in the long term, it will decline, whether or not it is maximizing "utility" and "perfect." We will take no position on "market failure" or "government failure." Instead, we will only measure the returns of systems with different levels of regulations in different conditions. All living systems regulate their internal environment. But the levels of regulation are system specific. We will not argue whether humans are "rational" or "irrational." Emotions often narrow the options in decision making. At the same time, they reduce cost in decision making. As long as certain emotion generates net benefits for its host, it will be preserved, whether it is "rational" or "irrational." In each case, we will assess the returns of the specific systems under specific conditions. Of course, we will make mistakes in assessments. But the ability to make falsifiable statements is the very hallmark of a scientific theory. A theory based on observable quantities can be subject to empirical testing and can improve from empirical testing. This is very different from the statement of "maximizing utility" in established economic theories. Whatever someone does, one can always argue that he is maximizing his utility by redefining his utility function in new ways.

Over time, people have developed great results on observable quantities in social sciences. But the potentials of these results are often underappreciated in an environment where utility is the main measure. For example, John Kelly developed a formula linking investor behavior and investment return more than half-century ago. His result has been applied successfully by many investors. However, his return-based theory is not compatible with the utility-based theory. Kelly's ideas were rejected by the academic establishment and have been largely ignored in academia. A detailed account of this history was presented in William Poundstone's fascinating book, *Fortune's Formula: The Untold Story of the Scientific Betting System That Beat the Casinos and Wall Street.* I struggled for many years to develop a mathematical theory to describe investors' behaviors. Only after I read Poundstone's book, I realized that my results are extensions of Kelly's theory. Had I known Kelly's theory earlier, it could have saved me from years of struggle. A major purpose of this book is to present many brilliant ideas by early pioneers in a unified framework. We hope that more people can access these ideas easily and do not have to waste tremendous amount of time struggling to redevelop the same ideas over and over again.

If it is so fruitful to study social issues from observable quantities, why the theoretical foundation of economics is still built on utility, an unobservable quantity? In his 1949 book, *Human Behavior and the Principle of Least Effort*, George Zipf advocated that social sciences should be built on observable quantities. He further pointed out that the elite will resist this idea and use the power of academic appointment to deter people from pursuing this approach. Instead of measuring the gains and losses of different parties in a particular situation, the elite like to declare that everyone maximizes their "utilities." Hence, the current situation is "optimal." We can and do make measurements on empirical data. But to be a respectable academic, one has to restrict his search of truth outside the domain dictated by political correctness. One can only study the patterns on observable quantities on minor issues. Indeed, the very purpose of political correctness is to suppress discussion on facts that will harm the interest of powerful groups. This is the most important reason why it is so difficult to make progress on fundamental issues in social studies.

Historically, the exchange of ideas between biology and economics has been very fruitful. Both Charles Darwin and Alfred Wallace were inspired by Thomas Malthus' population theory when they developed the theory of natural selection. Similarly, biology was considered "the Mecca of the economist." However, established economic theories are equilibrium theories, while biological systems are understood as non-equilibrium systems. This and many other reasons limited the knowledge flow between social and biological sciences. We will present a common platform for both social and biological systems. This will make it easier to apply insights from one area to another area. There are many advantages for an integrated approach to social and biological systems. Biological studies cover many more species over a much longer time period than social studies. Observations on other species are often more objective than observations on ourselves. Therefore, principles derived from biological studies tend to be more general and more robust than those from social studies. On the other hand, human societies are the most intensely studied biological group. Many ideas and mathematical techniques developed in economic theories can be applied to very general problems in life science. In the last several decades, our knowledge of biology has grown tremendously. This makes it very difficult for us to gain understanding on broad range of problems. A perspective from economics will weave many seemingly disparate facts into a coherent picture. This will greatly simplify our learning.

Many people have recognized that the standard economic theories are not built on a solid foundation (Hall and Klitgaard 2011). But they often have difficulty connecting basic scientific principles to specific economic policies and social structures. Our purpose is to introduce a common foundation for science and economics, so insights gained from science and economic theories can be applied to broader areas. This book is an attempt to reach a broad audience who are concerned about the current state of economic theory and the future of our society. In the first half of the book, we will present basic ideas and discuss policy issues without using mathematics. Instead, we rely heavily on intuition, which provides great under-standing on most important issues. Mathematical analyses are concentrated at the later part of the book. We provide detailed background information and discussion on historical developments of related mathematical theories to make it easier for the people to see the evolution of ideas and difficulties encountered by early pioneers.

Acknowledgments

Many colleagues, friends, and students have discussed ideas with me over various stages in the development of the theory. Jamie Galbraith is a great help. This book is dedicated to him.

David Packer, my editor from Springer guides me through the process of writing with great expertise. Charlie Hall and Kent Klitgaard went over the whole manuscript and made many improvements on the writing. Norm Jacob also went through part of the manuscript and helped me revise it.

Contents

Chapter 1
Major Factors in Biological and Social Systems

1.1 A Common Measure of Performance

For any biological or social system to survive and prosper, it has to provide a non-negative rate of return. In finance theory, the performance of a business is measured by its rate of return on monetary investment. In biological theory, the performance of an organism is measured by its rate of return on biological investment, or change of population size. In energy extraction, the rate of return is measured as energy return over energy invested.

However, neoclassical economists assume that human beings and social systems maximize utility instead of generating positive return. From our daily experience, when we are hungry, we try to obtain food; when we are thirsty, we try to obtain water; when we are sexually mature, we long for a mate. Our preferences and activities are indeed directed by our short term needs, or utilities. But short term needs and "utilities" are ways to achieve long term goals of positive rate of return. Over the long term, whether a system survives and prospers depends on its rate of return. A business will fail if it loses money over the long term, whether or not it maximizes its utility. An investor will have to exit the stock market if he loses all his money, whether or not he maximizes his utility. This is not to suggest any short term activities we perform will promote positive returns. We are all constrained by the structures of ourselves and the environment. We all have limited capacity to forecast and respond. After all, most species that appeared on the earth went extinct; most once powerful and prosperous societies eventually collapsed; most businesses that once existed closed down. As researchers of biological and social systems, we hope to understand why a system does well or fails.

Establishing an objective measure of performance does not mean the human mind and emotion are not important in understanding human societies. On the contrary, the human mind and emotion have evolved to generate positive biological returns. It is our love of children that enables us to spend tremendous amounts of effort to feed and protect helpless babies. It is because of the irresistible attraction to

© Springer Science+Business Media New York 2016
J. Chen, *The Unity of Science and Economics*,
DOI 10.1007/978-1-4939-3466-9_1

the opposite sex that two very different individuals live together to raise the next generation. This does not mean that our emotions always help us gain a positive return. The human mind is an evolutionary product of the past. The environment we live in today is very different from the past. What worked in the past may not work well in the future A return-based theory will allow us to analyze the long term impacts of our own behaviors, the strategies of our businesses, the structure of our societies, and the policies of our governments.

If a return based theory provides much clarity about our society, why the dominance of a utility based theory in economics? We can go back to the very first paper that introduced the concept of utility in economics. It was written by Daniel Bernoulli, originally published in 1738 (Bernoulli 1738). In that paper, he showed that the arithmetic rate of return does not provide a good measure of return and proposed to use the concept of utility to replace arithmetic rate of return. Daniel Bernoulli's utility function was equivalent to the geometric rate of return. The geometric rate of return provides a better measure of return than the arithmetic rate of return. If we adopt the geometric rate of return as the measure of return, we can resolve the problem raised by Bernoulli without resorting to the utility function.

By adopting a common measure of performance for both biological systems and social systems, we can understand biology and social sciences as an integrated theory. Social systems, like other biological systems, are enabled and constrained by physical resources and physical principles, and require non-negative return for their long term viability. There have been persistent suggestions that social theories should be built on the foundation of biology (Barkow et al. 1992). Yet it is often assumed that there is a fundamental difference between the two: genetic mutations are generally considered random while human activities are considered purposeful. However, genetic mutations and human activities are problems at different levels. Human beings evolve through genetic mutations as well and many animal activities are purposeful. Furthermore, more precise observation shows that genetic mutations are not completely random. When, where and how fast genes mutate is influenced by many environmental factors. The regulation in genetic and epigenetic changes in organisms is highly directed to enhance their survival under different kinds of environments. Since directed and informed changes often provide a higher rate of return than completely random ones, purposeful changes evolve both in social and biological systems. But sometimes random changes can help bring new and better results to systems stuck in local but not global optimums. For example, some optimization software adopts random changes periodically to test whether the obtained results are global optima. Because both purposeful and random activities are present in biological and social systems, we will not segregate the study of social systems from other biological systems.

An integrated theory of biology and human society enables us to understand better the long-term trends of human societies. After the rise of oil prices in the 1970s, many countries regained economic growth after deep recessions. Neoclassical economic theory contends that markets can overcome scarcity of resources. But from a biological perspective, any living systems, including human societies, require above replacement fertility to be long term viable. Fertility in most wealthy countries dropped below the replacement rate after the 1970s, rendering their rate of return on biological investment negative. The initial drop in the fertility rate reduced the number of dependent children. This reduced the cost of raising children. Many more adults, especially women, became available as workers. Countries in demographic transition often enjoy a high rate of growth in economic output for several decades. But eventually, as the majority of the working population becomes old with far fewer next generation workers to replace them, the ratio of working population to total population declines. Monetary activities will decline eventually. Biological theory foretold the social problems of the former Soviet blocks, Japan, Europe and many other countries several decades earlier, when their biological return turned very negative. We could have avoided these social problems if our policies are guided by biological theories.

Some people contend that human beings are social animals and biological studies focus on individuals. However, many species of animals are social animals and a great amount of research has been done on social animals. Furthermore, each multicellular organism, such as a human being, is a community of different cells with different characteristics and different locations. Different cells communicate and coordinate with each other to accomplish goals that are impossible to achieve by individual cells. We can gain deep insight about communication, coordination and regulation of different cells from studies in animal physiology.

Our goal is to understand what factors impact the long-term return of biological and social systems. In the next several sections, we will discuss some of the most important factors.

1.2 The Necessity of Fixed Costs

Thermodynamics is often called the economic theory of nature. From thermodynamic theory, useful works can be obtained only when a differential, or gradient, exists between two parts of a system. Plants can utilize sunlight to generate chemical energy because the sun is much hotter than the earth. We can utilize water flow to generate electricity when water flows from higher place to lower place. However, fixed investment is required before positive return can be generated. For example, plants need to make chlorophyll before they can transform solar energy into chemical energy. We need to build dams before electricity can be generated. In general, all organisms require a fixed set of genes before they can reproduce themselves.

In the language of economics, it requires fixed costs to transform resources to be used by all living organisms profitably. Specifically, fixed costs reduce variable costs. In general, a lower variable cost system requires higher fixed cost, although the reverse is not necessarily true. We will list several examples from engineering, biology and economics to illustrate the tradeoff between fixed and variable costs.

In electricity transmission, higher voltage will lower heat loss. But higher voltage transmission systems are more expensive to build and maintain. The differential of water levels above and below a hydro dam generates electricity. The higher the hydro dam, the more electricity can be generated. But a higher dam is more expensive to build and needs to withstand higher water pressure. In an internal combustion engine, the higher the temperature differential between the combustion chamber and the environment, the higher the efficiency in transforming heat into work. But it is more expensive to build a combustion chamber that can withstand higher temperature and pressure. Diesel burns at higher temperature than gasoline. This is why the energy efficiency of diesel engine is higher than that of gasoline engine and the cost of building a diesel engine is higher as well. In general, the higher the gradient, the more efficient energy can be transformed into useful work. This is the famed Carnot's Principle, the foundation of thermodynamics. At the same time, creating and maintaining high differential itself is physically difficult and hence financially expensive. The economic principle is consistent with, indeed, a consequence of the physical principle.

Warm blooded animals can generate high energy output longer than cold blooded animals because their bodies are maintained at high energy levels to ensure fast biochemical reactions. So warm blooded animals can control and consume more resources than cold blooded animals. But the basic metabolism rates of warm blooded animals are much higher than the cold blooded animals. Warm blooded animals have to eat much more than the cold blooded animals of the same weight to keep alive.

Shops located near high traffic flows generate high sales volume. But the rental costs in such locations are also higher. Well-trained employees work more effectively. But employee training is costly. People with higher education levels on average command higher income. But education takes time, effort and money.

The level of fixed cost of a system is often its defining characteristics. It is often used as a classification criterion in many research areas, although the term "fixed cost" is not necessarily used. In cultural study, cultures are often classified as high context cultures and low context cultures. In ecological study, species are classified as K species (high fixed cost) and r species (low fixed cost). In the study of social systems, societies are often classified as complex (high fixed cost) or simple (low fixed cost). Many debates in our societies are about fixed cost. When we say education is a right, we really mean that education should be part of the fixed cost of the society, that the state should pay for public education out of tax revenue.

For any living organism to be viable, its cost to obtain resources cannot exceed the value of the resources over its life cycle. Similarly, for a business to be viable in the long term its average cost of operation cannot exceed its average revenue. Costs include fixed cost and variable cost. The main feature of an organism or a business is the structure of its fixed and variable cost so its total cost can be lower than its revenue in specific environments. Historically, lower fixed cost systems often appear earlier than higher fixed cost systems, which often evolve from lower fixed cost systems. Single celled organisms appear earlier than multicellular organisms. Cold blooded animals appear earlier than warm blooded animals. Small family owned shops appear earlier than large global companies. Fixed costs in wealthy societies are also higher than fixed costs in poor societies. Wealthy societies often have compulsory secondary education and easy to access post-secondary education while people in poor societies often start working at an early age. Wealthy societies often have well maintained roads, free public libraries, relatively open and fair legal systems, which require high maintenance costs, while poor societies usually don't. Many people feel that the world will continue to evolve toward more complex, higher fixed cost systems, with only occasional setbacks. However, if we look at longer time spans and broader scales, the evolutionary patterns are subtler.

> The 'Doctrine of the Unspecialized'... describes the fact that the highly developed, or specialized types of one geological period have not been the parents of the types of succeeding periods, but that the descent has been derived from the less specialized of preceding ages.... The validity of this law is due to the fact that the specialized types of all periods have been generally incapable of adaptation to the changed conditions which characterized the advent of new periods.... Such changes have been often especially severe in their effects on species of large size, which required food in great quantities.... Animals of omnivorous food-habits would survive where those which required special foods would die. Species of small size would survive a scarcity of food, while large ones would perish. ... An extreme specialization...has been, like an overperfection of structure, unfavorable to survival. (Cope 1896, p. 173–174)

We often have the impression that higher fixed cost systems are better than lower fixed cost systems. For example, simple organisms or societies are generally called primitive while complex organisms or societies are generally called advanced. However, biologists now recognize that systems of different fixed costs are adapted to different kinds of environments. Indeed, all life forms, simple or complex, are successful outcomes of unbroken chains of almost four billion years of survival and reproduction. The proper amount of fixed cost a system needs to invest mainly depends on the amount of resources available. In general, in an environment of abundant resources, large, high fixed cost systems dominate; in an environment of scarce resources, small, low fixed cost systems breakeven easier. However, the amount of resources a system can access depends partly on its ability to control and utilize particular resources. In general, high fixed cost systems are more capable of controlling and utilizing resources than are low fixed cost systems. The net impact can only be measured by the return of a system.

In the last several hundred years, with the large scale use of fossil fuels, higher fixed cost social systems did well most of the time, expanding their systems globally. But in the past several decades, with continuous depletion of high quality resources and continuously rising living standards, fertility rates in most wealthy countries, as high fixed cost systems, have dropped below the replacement rate. This means that the biological returns in these places have already turned negative. To most economists, demographic changes are not the fundamental determinant of long term economic activities. So they could not foresee serious problems in the future. When economic recessions occur, neoclassical economists tend to view them as temporary breaks from the continuous growth. During the recent Great Recession, policymakers injected large amount of resources to pump up financial and auto sectors, two of the highest fixed cost industries. Policymakers expected these sectors generate high profits and large amounts of employment once the economy start to grow again. However, from the perspective of resource scarcity, the bail out of these high fixed cost industries only delayed the necessary adjustments toward a lower fixed cost society and made the whole society more vulnerable to future uncertainties.

We will look further into how the increase of fixed costs affects the rate of return. The increase of fixed cost is often associated with the division of labor. Instead of a single cell handling all biological functions, a multicellular organism often consists of many different organs, such as heart and liver, each responsible for specific functions. Unspecialized stem blood cells are capable to reproduce easily. But once these unspecialized stem cells become red blood cells and various white blood cells, they cannot reproduce themselves anymore. In general, specialized cells, such as red blood cells and neuron cells, often are infertile or less fertile. Similarly, in human societies, highly trained professionals, especially female professionals, are often less fertile. Being a parent requires a lot of general skills, such as changing diapers, feeding the babies and cooking for the family, that are possessed by so called unskilled workers but are lost or degenerated among many highly trained specialists. If the fixed cost of a system becomes too high and many people in the system become over specialized, the average fertility of the social system will drop below the replacement rate, rendering the system unsustainable.

The tradeoff between fixed cost and variable cost is universal in economic activities. However, this tradeoff is often not explicitly discussed in the same literature and often not considered in policy issues. For example, electricity generated from solar panel is considered clean energy because the solar panel does not need fuels that will cause environmental problems. But the manufacturing of solar panels is highly resource intensive and highly pollutive. However, the pollution from manufacturing solar panels, the fixed cost part of the solar electricity, is rarely mentioned in policy discussion. While it is in the interest of the promoters of "clean" energy and "renewable" energy to avoid discussing such issues, a good economic theory should provide guidance to understand the big pictures.

The necessity of fixed cost has broad implications. It takes time to recoup the initial investment. Therefore it is a fundamental feature for organisms and investment projects to have lifespans or durations. How lifespans of different organisms are evolved? How project durations are determined? Are long life systems necessarily better? Uncertainty is an unavoidable part of nature. But the same level of uncertainty will have different impacts on systems with different fixed costs and durations. How they are related? The fixed costs are incurred or committed at early stages of organisms or projects' life. But the expected payoffs, which occur in the future, are only estimates. Therefore expected payoffs from investments are subject to discounting. How should discount rates be determined? What are the consequences of different level of discount rates? When market size is large, high fixed cost systems, with low variable cost, will generate high rate of return. Are larger market sizes always better? We will explore these and other questions in the next several sections.

1.3 Lifespans of Organisms and Investment Projects

How do lifespans or durations of organisms and projects affect the rate of return on our investments? If the duration of a project is too short, we may not be able to recoup the fixed cost invested in the project. If the duration of a project is too long, the variable cost may become too high and the rate of return will turn negative. For example, if a taxi driver keeps his car too long, the maintenance cost of the car may become very high and the car may break down very often. As a result, the total cost may become higher than the revenue. This is why individual life does not go on forever. Instead, it is of higher rate of return for animals to have finite life spans and produce offspring. In social sciences and policy discussion, longer life span is often used as an indicator of higher quality of a social system. However, societies that enjoy a long life span, such as Japan, often struggle with below replacement fertility. We will analyze the relation among fixed cost, lifespan or project duration, and rate of return in greater detail.

When the level of fixed cost increases, it often takes longer time for a project to breakeven. Large animals and large projects, which have higher fixed cost, often have longer lifespan. There is an empirical regularity that animals of larger sizes generally live longer (Whitfield 2006). The relation between fixed cost and duration can be also applied to human relation. In child bearing, women spend much more effort than men. On average women value long-term relationships while men often seek shorter- term relationships (Pinker 1997).

When the duration of a project keeps increasing, variable cost will keep increasing and the return of a project will eventually turn negative. Hence, duration of a project or an organism cannot become infinite. For life to continue, there has to be a systematic ways to generate new organisms from old organisms. From earlier discussion, for a system to have a positive return, fixed assets have to be invested first. Thus old generations have to transfer part of their resources to younger generations as the seed capital before younger generations can generate positive return. Therefore, there is a universal necessity of resource transfer from one generation to the next generation in biological and social systems. "Higher" animals, such as mammals, generally provide more investment to each child than "lower" animals, such as fish. In human societies, parents provide their children for some years before they become financial independent. In general, wealthy societies provide more investment to children before they start to compete in the market than poor societies. In businesses, new projects are heavily subsidized at their beginning stages by cash flows from profitable mature projects.

Empirical evidence exhibits the inverse relationship between lifespan and fertility. Lane (2002) provided a detailed discussion about the tradeoff between longevity and fecundity in the biological systems.

Notwithstanding difficulties in specifying the maximum lifespan and reproductive potential of animals in the wild, or even in zoos, the answer is an unequivocal yes. With a few exceptions, usually explicable by particular circumstances, there is indeed a strong inverse relationship between fecundity and maximum lifespan. Mice, for example, start breeding at about six weeks old, produce many litters a year, and live for about three years. Domestic cats start breeding at about one year, produce two or three litters annually, and live for about 15 to 20 years. Herbivores usually have one offspring a year and live for 30 to 40 years. The implication is that high fecundity has a cost in terms of survival, and conversely, that investing in long-term survival reduces fecundity.

Do factors that increase lifespan decrease fecundity? There are number of indications that they do. Calorie restriction, for example, in which animals are fed a balanced low-calorie diet, usually increase maximum life span by 30 to 50 per cent, and lower fecundity during the period of dietary restriction. … The rationale in the wild seems clear enough: if food is scarce, unrestrained breeding would threaten the lives of parents as well as offspring. Calorie restriction simulates mild starvation and increase stress-resistance in general. Animals that survive the famine are restored to normal fecundity in times of plenty. But then, if the evolved response to famine is to put life on hold until times of plenty, we would expect to find an inverse relationship between fecundity and survival. (Lane 2002, p. 229)

Lane went on to provide many more examples on the inverse relation between longevity and fecundity.

The necessity of the fixed cost investment and the finiteness of lifespan determine that resource transfer from old generation to new generation is essential for the long term viability of a system. However the process of resource distribution is often the source of many conflicts between generations and within the members of the same generation. Each child wants more resources from parents. But parents prefer children to become independent early so resources can be distributed to younger or unborn children. Old mature industries, which need little R&D expense, prefer low tax systems. But young high tech industries, which rely heavily on universities to provide new technologies, employees and users, strongly advocate government support in new technologies. Businesses prefer lower tax rates. But educational institutions, which mainly train the younger generation and receive much of their incomes from governments, prefer higher government revenues.

The conflict between generations often starts with pregnancy. "The greater the amount taken by the fetus, the greater its birth weight, but the less its mother would have for other purposes. Lighter babies would have had a reduced probability of survival, but costly pregnancies would have increased the mother's vulnerability to disease, reduced her ability to care for existing children, and decreased her chances of reproduction again (Haig 1993, p. 496)."

In our daily life, we all experience conflicts between generations. When contraception technology is available, many people decide to delay raising children to delay the transfer of resources to the next generation. With aging parents, more and more children are born with genetic defects. In many societies, the fertility rates also drop below the replacement rate because of the delaying or avoidance of resource transfer to the next generation.

The increase of retirement age is often advocated as the solution for potential labor shortage in societies with low fertility rate. But many countries with low fertility rates have very high youth unemployment rates. Younger people are more productive in most works. Economic performance will improve if more young people can have jobs. In general, systems with below replacement rate fertility need to shorten the duration of the work period to restore the viability of the systems. If older people can retire earlier, more young people can move out of unemployment. With financial security, young people in their prime reproductive age can have more children. However, in a society with low fertility and long lifespan, the proportion of seniors will be very high. Furthermore, young and middle aged people will grow old but old people will not turn young. Resources allocated to seniors generally get less resistance than resources allocated to young people. Since unborn and small children have little political clout, their interests are often unrepresented in public discussion. Empirical evidence shows that when fertility rates in a society drop below replacement rate for several decades, it is very difficult for the system to return to the replacement rate.

1.4   Uncertainty

Uncertainty is an integral part to all living systems. But different systems respond to uncertainty in different ways. In general, simple low fixed cost, short duration systems are quicker to adapt to environmental changes than complex high fixed cost, long duration systems. However, high fixed cost systems often possess more resources to work with. They often spend more resources to maintain stability in internal and external environment to reduce uncertainty. They also evolve mechanisms to adjust themselves periodically. We will use some examples to illustrate the strategies adopted by systems with different fixed costs.

Lower fixed cost systems in general have shorter life span than higher fixed cost system. The mutation rates or the rates of change of lower fixed cost systems are higher. This gives lower fixed cost systems advantages in initiating and adapting changes. For example, AIDS virus is much smaller than human beings and can mutate much faster. This makes it difficult for humans to develop natural immune response or develop drugs to fight against AIDS virus. However, higher animals develop a general strategy in immune systems that has been very effective most of the time. Instead of developing one kind of antibody, our immune systems produce millions of different types of antibodies. It is highly likely that for any kind of bacteria or viruses, there is a suitable type of antibody to destroy them. This strategy is very effective but very expensive, because our body needs to produce many different kinds of antibodies that are useless most of the time. When we are too young, too old, too weak, or too stressed, our bodies don't have enough energy to produce large amount of antibodies. That is when we become vulnerable to infection. Many empirical studies have shown that when challenged with pathogens organisms divert energy to combating the invader, but at an overall cost in energy to the organism, leaving less energy available for other contingencies.

Other than adapting to the ever changing environment, organisms also regulate internal and sometimes external environment. In general, higher fixed cost systems spend more resources to regulate environment to reduce the level of uncertainty. Human beings, as warm blooded animals, regulate our body temperature around

37 degrees, regardless of external temperature. The constancy of body temperature greatly reduces the uncertainty for our body's many chemical and physical processes. However, the maintenance of constant temperature is extremely energy intensive. Warm-blooded animals require a lot more food intake than cold blooded animals of the same size. When they cannot find any food for several days, their body functions weaken seriously. This causes great increase in uncertainty.

Living systems not only regulate internal environment, they also regulate external environment. Among all animals, human beings have the greatest capacity to modify our external environment. We clear land for agriculture, which provides much more certain supply of foods than hunting and gathering. We build houses to shelter us from the uncertainty brought by wind, rain and snow. In wealthy societies, we take many measures to reduce uncertainty in many aspects of life (Galbraith 1958). When we lose jobs, we receive unemployment compensation. When we are disabled, we get disability assistance. Our buildings are cooled in summered and heated in winter to maintain a constant indoor temperature. Water is chlorinated to eliminate bacteria. Populations are vaccinated to reduce infectious diseases. Many measures are taken to make life as predictable and as pleasant as possible. In general, uncertainty in wealthy countries is very low compared with poor countries. In environment with low uncertainty, investment with high fixed cost and long duration often generate high rates of return. But measures to reduce uncertainty require resources. For example, to keep buildings warm in winter, we need natural gas to provide heat.

It is more difficult for high fixed cost systems to change. But high fixed cost organisms do develop mechanisms to change periodically to adapt to changing environment. Most complex animals reproduce sexually. Sexual reproduction is very expensive compared with asexual reproduction. Just imagine how much effort we spend on relationship and how difficult relationship can become. However, by reproduce sexually, the genes of our offspring are reshuffled substantially. The genetic diversity of our offspring makes some of them more adaptable to the changing environment. The genetic changes also make parasites in our bodies less able to get established on our offspring. In social systems, democratic countries have elections periodically to change the leaders of the systems.

In general, mature industries with low levels of uncertainty are dominated by large established companies. For example, household supply industry is dominated by P&G; soft drink industry is dominated by Coca Cola and Pepsi Cola. Fast changing industries, such as IT, are pioneered by small and new firms. Microsoft, Apple, Yahoo, Google, Facebook and many other successful businesses are started by one or two individuals and not by established firms with capital and expertise, although many of these firms are currently establishing themselves as oligopolies. Similarly, in scientific research, mature areas are generally dominated by top researchers from elite schools, while fundamental ideas are often initiated by newcomers or outsiders. Most economists receive very narrow training in the neoclassical perspective, often to the exclusion of physical and biological sciences. But some of the most innovative thinkers in economics are outsiders. Neoclassical economics, the dominant economic theory today, was founded around 1870 by

Jevons, Walras and others. Both Jevons and Walras, considered by many as the greatest economists in the nineteenth century, were trained in science and engineering. In the last several decades, some of the great yet under-appreciated insights about economic activities, were initiated by pioneers trained in science.

The relations between fixed cost, uncertainty and duration apply to biological systems as well. We will compare the properties of RNA and DNA molecules. DNA is made from RNA with an extra step of chemical reaction. So DNA is more costly to make than RNA. DNA molecules are more stable and last longer than RNA molecules. So DNA is the preferred molecule to carry genetic information for most organisms. The only organisms to use RNA to carry genetic information are some viruses, which are very small and mutate very fast. For example, AIDS viruses use RNA to carry genetic codes. Any larger organisms, from bacteria on, use DNA to carry genetic codes. But RNA molecules are used in our bodies for many functions that do not require high level of precision and do not last very long, for RNA molecules are cheaper to make than DNA. The chemical properties and biological functions of DNA and RNA show that fixed cost, uncertainty and duration are intimately correlated in biological systems. They also show that economic properties of living systems, including human societies, are ultimately derived from physical and chemical properties of particles.

1.5   Discount Rate

From monetary policies to the climate change problem, from the burden of private credit card debts to the evaluation of public projects, discount rate is the central issue. Much effort has been made to understand the tradeoffs among investment and consumption behaviors at different time periods, yet there is little clear understanding about the nature of discounting. Martin Weitzman, the world's leading expert on the social discount rate, commented:

> The concept of a "discount rate" is central to economic analysis … Because of this centrality, the choice of an appropriate discount rate is one of the most critical problems in all of economics. And yet, to be perfectly honest, a great deal of uncertainty beclouds this very issue. … The most critical single problem with discounting future benefits and cost is that no consensus now exists, or for that matter has ever existed, about what actual rate of interest to use. (Weitzman 2001, p. 260)

The main problem in the theory of discounting, as pointed out by Weitzman, is that "an economist who knows the literature well" is "able to justify *any* reasonable social discount rate by some internally consistent story". We will discuss how the discount rate is related to other factors in biological and social production systems, such as fixed cost in production, duration of production or life span, and uncertainty. The relations among different factors in a production system will put constraints on the ranges of discount rate that are viable in particular environments. These constraints will help us understand how discounting should be applied in different situations. But before our analysis, we would like to list some puzzles related to discount rates.

The borrowing rates for banks are very low. But the credit card interest rates that banks can charge their customers are very high. How can the large interest rate differential be maintained over a long time? From another perspective, individuals can obtain a line of credit at a much lower interest rate than the credit card interest rate. Why do so many people still maintain a large amount of credit card debt, without replacing it with a line of credit?

Most economists are staunch proponents of the efficiency of markets, especially when the market is very liquid and transparent. The short term money market is among the most liquid and transparent markets in the world. Yet, most economists support that the short term discount rate, possibly the most important factor affecting economic performance, should be determined by a small group of "independent" professionals from central banks. Does that mean the market is only capable of being efficient on minor issues and not on major issues? To answer this question we turn to yield curves.

In general, yield curves slope upward. Loans with longer maturity pay higher interest rates than loans with shorter maturity. At the same time, empirical evidence suggests that humans discount the long term future at lower rates than the short term future (Ainslie 1992; Berns et al. 2007). Many policy papers also advocate discounting long term projects at lower rates than short term projects (Weitzman 2001; Newell and Pizer 2003). Why do market discount rates and psychological and policy discount rates move in different directions in relation to the increase in project duration?

In the following, we will discuss how discount rate is related to other main factors in production. This will help resolve the puzzles. First, discount rates are closely related to fixed cost in production. When we have invested a large sum on something, we will take extra care that the value of investment depreciates slowly. From another perspective, in a low interest rate environment, the cost of borrowing is low. Investments with higher fixed cost will benefit. Investments with lower fixed costs are less sensitive to the level of discount rates. Human beings instinctively understand the relation between fixed cost and discount rate, although most of us are unaware of this. Many psychological experiments show that the "rate of temporal discounting decreases with the amount of reward" (Thaler 1981; Green et al. 1997). In the field of human psychology, this empirical regularity is called the "magnitude effect" (small outcomes are discounted more than large ones).

An earlier work by Ainslie and Herrnstein (1981) provides similar understanding:

> The biological value of a low discount rate is limited by its requiring the organism to detect which one of all the events occurring over a preceding period of hours or days led to a particular reinforcer. As the discounting rate falls, the informational load increases. Without substantial discounting, a reinforcer would act with nearly full force not only on the behaviors that immediately preceded it, but also on those that had been emitted in past hours or days. The task of factoring out which behaviors had actually led to reward could exceed the information processing capacity of a species.

Differences in fixed costs in child bearing between women and men would affect the differences in discount rates between them. Women spend nearly all the effort in child bearing. The high fixed investment women put in child bearing would make women's discount rate lower than men's. An informal survey conducted in a classroom survey showed that discount rates of the female students are lower than that of the male students.

In poor countries, lending rates are very high; in wealthy countries, lending rates charged by regular financial institutions, other than unsecured personal loans, such as credit card debts, are generally very low. To maintain a low level of lending rates, many credit and legal agencies are needed to inform and enforce, which is very costly. Wealthy countries are of high fixed cost, from education to social programs to infrastructures. So they are willing to put up the high cost of credit and legal agencies because the efficiency gain from lower lending rate is higher in high fixed cost systems. In the last several hundred years, there has been in general an upward trend in living standard worldwide. There has also been a downward trend in interest rates (Newell and Pizer 2003).

Second, discount rates are closely related to the duration of production or lifespan of organisms. In steady states, the rate of reproduction is equal to the rate of death. Therefore, in biological literature, the discount rate is often set to be equal to the rate of reproduction (Stearns 1992). Bacteria can reproduce themselves in just thirty minutes. Humans can reproduce only after ten or more years old. Hence, discount rates for bacteria are measured in hours while discount rates for humans are measured in years. In general, organisms with longer life spans have lower fertility rates and hence they have lower discount rates (Lane 2002). Many empirical studies have documented that humans, as well as other animals, often discount long duration events at lower rates than short duration events (Frederick et al. 2004). This pattern is called hyperbolic discounting.

Third, discount rates are closely related to uncertainty. In general, organisms facing a high level of environmental uncertainty, such as predation, have higher discount rates (Stearns 1992). In economic theory, the standard quantitative model on the relation between discounting and risk is the capital asset pricing model (CAPM), which states that systems with higher risk is discounted at higher rate. The same idea about the relationship between discount rate and uncertainty was reached in the study of human psychology. "The same discount curve that is optimally steep for an organism's intelligence in a poorly predictable environment will make him unnecessarily shortsighted in a more predictable environment (Ainslie 1992, p. 86)".

The above discussion shows that discount rates, fixed costs, duration of production and uncertainty are highly related. Lower discount rates are positively correlated with high fixed costs, long duration, and low uncertainty. High discount rates are positively correlated with low fixed cost, short duration, and high uncertainty. These results are consistent with the classification of organisms in ecological studies. Organisms are often classified as K strategists and r strategists.

K strategists often have high fixed costs, long duration and low discount rates. They are highly competitive in stable environment. r strategists often have low fixed costs, short durations and high discount rates. They thrive in volatile environment (MacArthur and Wilson 1967). Our results about the relation between discount rate and other factors are very similar to ones that are obtained in the study of mind of humans and other animals (Ainslie 1992). This shows that the human and animal mind, a product of evolution, understands the important relations in life very well.

In most of our evolutionary past, uncertainty was high, lifespan was short, fixed investment was low. So our discount rate is high. Many of us are willing to borrow at high interest rate. They feel that high interest rate credit card debt is normal. So they accumulate credit card debts casually and don't put much effort in looking for lower rate alternatives, such as line of credit. The accumulation of debts often has tragic consequences to individuals, their families and the society as a whole. This is probably why in many cultures and religions, money lending businesses are discouraged or banned. However, modern societies have developed technologies to extract abundant fossil fuels and other resources. With abundant resources and hence high economic output, high fixed cost investments generate high rate of returns. Low interest rate environment will stimulate high fixed investment. But interest rates will be high in a free financial market because the discount rate in our mind is high. In many countries, governments take over the task of setting short term interest rate to keep interest rate low. By maintaining low interest rates, governments help stimulate high fixed cost investment and overall economic growth. This is why short term interest rates are determined by governments and not markets. For mainstream economists, free market provides the optimal result. Any government action requires explanation from externality or imperfection, which are often arbitrary. But in the real world things are different. Once we recognize the relations among major factors in biological and social systems, it is easy to understand why governments, not markets, are setting short term interest rates in many countries.

There is another puzzle related to interest rate. Why consumers are allowed to accumulate large amount of credit card debts, which charge very high interest rate? Mainstream economists argue that free market provides optimal results. Hence the issue of credit card borrowing should be left to individuals. If this is so, why do legislations force people to save for retirement funds, which usually generate very low rate of return? If people could use pension money to pay off credit card debt, the total interest payment will be greatly reduced. At a theoretical level, why different logics are used to explain different things? But from the perspective of returns for financial institutions, it is easy to understand. Financial institutions generate huge revenues from both credit card businesses and pension fund management. Naturally, they will support both types of businesses. Logical inconsistency is unimportant compared with profit. This is another reason why a return based theory provides more explanatory power than other approaches.

1.6   The Volume of Output, Market Size and Abundance of Resources

The increase of market size and volume of output often increases profit or reduces average cost. However, average costs do not always decline with increasing outputs. Both economies of scale and increasing marginal costs play a role in overall costs. In educational system, primary schools are usually smaller than secondary schools, which in turn are smaller than universities. It would be difficult for primary school students to walk for a long distance to go to school. So a primary school caters to a very small market. Secondary school students can walk for longer distance but usually live with their parents. So a secondary school caters to mostly local students. University students can live independently, often away from home. So a university caters to local and distant students. As a result, the market size of a university can be larger. Market size of an organization is closely related to its fixed cost. Primary school teachers teach multiple subjects. A small primary school only needs one teacher for each grade. A secondary school needs specialized teachers for English, mathematics, science, physical educations and other subjects. A university usually has many specialized programs. Each program often has several specialized fields that need different faculty members. For example, a business school generally has department of accounting, finance, marketing and other programs. With increasing market sizes from primary school to university, their corresponding fixed costs increase as well.

Many factors determine the market size of a product or the volume of output of a business. An important factor is the level of resource abundance, especially energy abundance. When resources are abundant, more resources can be transformed into products people need and transportation cost is low in monetary term. People will buy a lot of different things in large quantity. When many large oil fields were discovered in late 20s and early 30s in the last century, oil prices dropped sharply. The rapid increase of oil supply derailed the coal based economy. This was partially responsible for the severity and length of the Great Depression. After the long and difficult adjustment period, many new cars were built and many new roads were paved to take advantage of the abundant oil. Many shopping malls and supermarkets with ample free parking spaces were built. They provided large variety of merchandise and served customers within driving distances. Gradually they replaced general stores, and downtown shopping districts which provided smaller quantities of essential goods and served customers within walking distance, as the main destinations for shopping. Low transportation cost also greatly increases the size of international trade, making the global economic activities more integrated. Most economists believe economic activities will become more integrated in the future. However, the productions of many crucial resources, such as oil, are near their peak or already peaked. So it is highly likely that trend of globalization and increasing market size may reverse soon. Historically, free trade zones have been growing and declining alternatively. Volume of economic output has been growing and declining over different time as well. Some people have already prepared and practiced various forms of localized economy, such as growing and eating local foods.

The market size for commercial goods has been expanding continuously for a long time. The same is true for the market sizes of languages. Because of the decreasing cost of transportation and communication, we tend to interact with more and different people. The benefit of using major languages, especially English, increases. Many languages with only a small number of speakers have disappeared or are disappearing. However, the market size for the states, the most important social units, has been in decline for some time. At least since the end of the World War II, the number of countries has been growing steadily. This means the average market size of the states, measured by area, has been declining since then. Soon after the end of the World War II, many Asian and African regions became independent countries. In the nineties of the last century, the Soviet Union and Yugoslavia split into several different countries. From time to time, some countries, such as Czechoslovakia, Sudan and Ethiopia, split into two countries. The expansion of free market and free trade often intensify the social division, making societies less stable (Chua 2003). It is likely that the number of countries will continue to increase in the future.

Market size is closely linked to fixed cost of a system. When fixed costs are low, an area may contain many independent units of production. But with higher fixed cost and increasing division of labor, fewer integrated systems can live in the same area. For example, a single grizzly bear, which is composed of trillion cells, require a large piece of land to survive. On the same piece of land, trillions of single celled organisms live there. The grizzly bear is very powerful, much more powerful than any single celled bacteria. But when any organ of the grizzly bear becomes weak, most of the trillion cells suffer as well. When the kidney cannot filter out waste effectively, all the trillion cells on the body of the grizzly bear have to live in a toxic environment. When the heart fails, all the trillion cells on the body of the grizzly bear die together. On the other hand, even many single celled bacteria die constantly, other bacteria can survive independently and the community of bacteria can prosper.

There is a parallel in human societies. In a simple society, families or tribes are often the independent units of production. While each family or tribe struggle mightily to stay afloat, the lives of different families or tribes often are not highly integrated. But in a high fixed cost system with extensive division of labor, most members of the society are integrated into the system. Soviet Union was once a splendid highly integrated system. It did so well in its early days that many people were fearful of its dominance in the near future. But a high fixed cost system is difficult to adapt. In a high fixed cost system, the deficiency of a small part will crumple the whole system. Over time, such deficiency will surely develop and spread, ultimately destroying the whole system. While the capitalist system is supposed to operate on private ownership, everyone and every business is obliged to pay tax. It means that everyone and every business is partly owned by the government. When the tax rate is high, the percentage of government ownership is high and our social system becomes highly integrated. In such systems, the freedom and right of each member is much less than what we would like to believe. For example, during the recent financial crisis, wealthy financial institutions are bailed out with taxpayers' resources, while ordinary people have no say in the decision making process. When the collective resources are forced to support the financial industry whenever the collapse of financial system seems imminent, the whole society becomes a single large market size, high fixed cost system, whether we like it or not. If everyone is forced to live in a single large building, the collapse of the building will bury us all.

Some biological processes become easy to understand from the sizes of output. There are twenty types of amino acids. Glutamine is often the first amino acid to be synthesized. But glutamine is sometimes converted to arginine before amino acids are transported to somewhere else (Willey et al. 2011, p. 703). It may be difficult to understand why organisms take the extra step in performing a task. But each arginine molecule contains four nitrogen atoms, the most nitrogen atoms for any amino acids. Transporting one arginine molecule carries four times as many nitrogen atoms as glutamine. Similarly, a hemoglobin molecule can carry four oxygen molecules while a myoglobin molecule only carries one oxygen molecule. Hemoglobin molecules transport oxygen over long distance and myoglobin molecules deliver oxygen to nearby cells. By carrying four oxygen molecules, hemoglobin reduces the cost of oxygen transportation. By carrying one oxygen molecule, myoglobin supplies the oxygen need for nearby cells. The amount of fixed cost in each investment depends on the size of the output.

1.7 On Demand and Supply

Demand and supply are two basic concepts in biology and economics. Biologists and ecologists recognize that conditions of an ecosystem are ultimately determined by the supply of resources, such as size of land, amount of water, sunlight, carbon dioxide and other nutrients. However, most Keynesian economists blame the lack of demand as the cause of poor economic conditions. They often prescribe policies to stimulate demand in times of economic downturn, while at the same time lamenting the high debts accumulated by the general public. Why people from different disciplines have different emphasis on demand and supply? We will apply the concepts discussed in earlier sections to analyze their arguments.

An economic system, like a biotic system, requires fixed cost investment. A system with higher fixed cost generally needs higher output to breakeven. At the same time, a higher fixed cost system, with lower variable cost, has the possibility of generating higher rate of return when the output is high. In a mostly growing economy, anticipation of larger market size encourages higher fixed cost investment. However, the initial market size may not be able to support the high fixed cost investment. If the expected market growth doesn't materialize, the investment will not be able to breakeven and face the risk of collapse. So it is essential to maintain high level of demand to keep high fixed cost investment profitable. Very often, the demand is maintained through borrowing. Since the economy is on the upward trend, borrowing can generally be repaid in the next cycle of economic boom. In a rising economy, stimulating demand will encourage high fixed cost investment, which provides high rate of return in a large market. Since modern economic theory is an adaptive product in a growing economy, most economists emphasize the demand side of the equation.

Biologists and ecologists usually consider more general patterns over long terms. From time to time, some biological systems find new ways to obtain resources to expand their supply. When resources are abundant, organisms multiply rapidly to consume more resources. Over the longer term, organism will occupy all available ecological niches. In the end, the growth of any biological system is constrained by resource supply. Because of technological developments to utilize fossil fuels and other resources in great scale, economic growth and population growth have been the norm for the last several centuries. However, in most wealthy countries, where per capita resource consumption is high, fertility rates have already dropped below the replacement rate. This suggests that supply of resources has already become the main constraint of the economic activities, at least in high resource consumption societies.

Politically, it is more attractive to promote the increase of demand. However, many households and governments are already heavily indebted. The amount of debt has been increasing over time. The problem with current economic system is not the lack of demand. Rather the problem lies at the high fixed cost of our economic system in wealthy countries, which requires high level of demand to breakeven. If the fixed cost of our social system becomes lower, we can generate positive return and make the system viable with lower level of demand. However, with the continuous economic growth of several centuries, we are getting used to the steady increase of spending, or demand. It is very difficult for us to accept the reality that continuous growth is never the normal state in any biological system, including human societies. Living on a constant solar input is very different from the exceptional circumstances of exploiting, for a few centuries, the stored sunlight of fossil fuels.

1.8 Concluding Remarks

According to established neoclassical economic theories, human beings are assumed to maximize utility. Since utility is subjective, maximizing utility can be many things to many people. The establishment economic theories also assume that market is the most efficient way to organize economic activities. But due to "imperfection" or "externality", most economists believe that regulations are needed from time to time "to make the market work". Education should be promoted by the governments, because education generates positive "externalities". Short-term interest rates are set by the central banks, because markets can be "myopic". These are rationalizations, not explanations. They don't tell us how much government should spend on education or what levels interest rates should be at.

In this chapter, I proposed the rate of return as the common measure of all biological and social systems and discussed the major factors that affect the rate of returns. Currently, most high resource consumption societies have negative biological rate of return, rendering these systems unsustainable. For these systems, the necessary reduction of resource consumption will make these systems viable again. Specifically, for these systems, fixed costs shall be lower, duration of investment shall be shorter, less resource shall be spent to reduce uncertainty, discount rate shall be higher and market size shall be smaller. In practice, reducing the tax rate, which will reduce government revenue, will achieve most of these goals. With lower government revenues, the sizes of governments, as the highest fixed cost systems, will be reduced; number of years of mandatory education will be reduced; average retirement age will be earlier; entitlement programs will be scaled back; governments and central banks will have less influence on interest rate and financial markets; supranational organizations, such as the Eurozone, will be abolished or scaled down, making each individual state more flexible in dealing with their own problems. This sounds like contemporary arguments of the political right. However, it is not. Medicine is the highest paid profession. It is also the profession most heavily subsidized by the government.

Among the major factors in social systems, the change of fixed cost often leads to corresponding changes in other factors. However it is often difficult to reduce fixed cost in an economic or social system, for several reasons. First, higher levels of fixed cost are widely associated with progress. Established economic theories generally emphasize the advantages of division of labor and of specialization, which tend to increase fixed costs in economic systems. The advantages of higher fixed cost systems are certainly real in an expanding environment while resources are cheap and abundant. However, over a longer time horizon, resource constraint is often more pronounced. Jesus once said, "Blessed are the meek, for they shall inherit the earth." This means lower fixed cost systems often have advantages over long term.

Second, changing the structure of an economic system is very costly and disruptive. It is something not likely to be undertaken while there is doubt about the long-range outlook. If the decline or difficulties are thought to be short term, it is often better off to reduce output while maintaining the essential production system intact; only when the decline is known to be long term does incurring the cost of change make sense. For this reason, strategists of denial about the long-term character of change in the resource environment also argue powerfully against making changes in the structure of fixed costs.

Third, reducing fixed cost often involves cutting the number and the incomes of people at high income levels. For example, the trial of Robert Pickton, a serial killer from the Vancouver area, cost the government over one hundred million dollars. Most of the money goes to highly paid professionals, such as lawyers. If the tax rate is lowered, tax income will be reduced. Governments won't be able to support as many highly paid professionals, who are politically influential. This is obviously difficult.

The reduction of fixed cost, with the corresponding increase of variable cost, will make our daily life less convenient and a great deal of marketing is the selling of convenience. The reduction of resource consumption will reduce our living standard. Hence it is often difficult to adopt policies that will reduce the fixed cost of a society. So far, it is mainly through the natural selection instead of adaptation that low fixed cost communities spread out. In biological systems, low fixed cost systems have higher fertility than high fixed cost systems. The same is true in human societies (Rushton 1996). Low fixed cost communities generally have higher fertility rates than high fixed cost communities. With currently low mortality rates among both high fixed cost and low fixed cost communities, low fixed cost communities spread out with respect to high fixed cost communities. Both selection and adaptation are at work in the evolution of human societies, like any other biological systems (Jablonka and Lamb 2006; Cochran and Harpending 2009).

In general, human beings, as a very successful evolutionary product on the earth, know what will benefit us. If a policy has both short term and long term benefits for most of us, it is unlikely that it has not been implemented. If a policy with significant long term benefit to the majority of people is not adopted, it is almost certain that the policy has negative short term impacts to many of us. Most of us understand the long term harm caused by drug use. But many people choose to use drugs because of the short term pleasure they bring. Short-term benefits of many social policies are reaped by members of the current generation who can influence decision making, while long term costs are paid by youth, children and future generations, who have little political influence. This makes it difficult to abandon such policies. Inevitably most of the policy recommendations derived from this economic theory will be politically incorrect or politically unpopular.

The established economic theories often assume technology development will resolve the problem of resource scarcity. New technologies often enable us to utilize resources that previously could not be tapped. At the same time, technology itself requires resource inputs. The net output of resources from new technologies may or may not be positive. In the next chapter, we will discuss in greater detail on the relation between resources and technologies.

This chapter provides an intuitive introduction to the relations among major factors in economic systems. We will present a mathematical theory about these relations in Chap. 3.

Chapter 2
Resource and Technology

2.1 The Importance of Natural Resources

The standard economic theory states that natural resources are only one factor in economic activities, which can be easily substituted by other factors, and output can be augmented by increases in technology. The depletion of natural resources is of little concern since for neoclassical economists for technological advance and resource substitution will generally overcome the problems caused by depleted resources. The standard economic theory is taught in almost all universities worldwide. It has a great influence on public opinion.

To understand better the relation between resources and technology, we will examine regions where the resource base is very narrow so the impact of resource depletion can be assessed more clearly. Sunlight is the most universal resource to the world. Many other resources, such as fresh water and fertile land, are derived from abundant sunlight. We will examine population change in a mining town in the North, where solar energy, the most important and universal natural resource, is scarce. Suppose in one mining town, there are 10,000 residents, of which 20 % are miners or mining related service providers. The remaining 80 % are policemen, teachers, doctors, pastors, bakers, grocery store cashiers and other service providers. Can we conclude, from this figure, that only 2000 people in the town depend on the mining activities? Suppose, after some years, the mine is exhausted. Can new technologies provide additional job opportunities or at least support the non-mining population? Some historical examples will offer a hint. In its heyday of gold rush, Barkerville, in northern British Columbia, Canada had over 10,000 residents. It was once the largest town in the west of North America. Its population dwindled to zero when the gold mines were exhausted. This is not an isolated example. In almost all mining towns in the north, once mines are exhausted, the towns become ghost towns. Any region requires resources to support its residents.

When resource bases are narrow, it is easy to recognize natural resources as the ultimate source of all economic activities and technologies as means to utilize

© Springer Science+Business Media New York 2016
J. Chen, *The Unity of Science and Economics*,
DOI 10.1007/978-1-4939-3466-9_2

resources. But often there are many different kinds of natural resources in the same place. People will move on to other natural resources after the depletion of one natural resource. For example, with the depletion of gold mines in California, people moved on to agriculture and other activities that require different kinds of resources, such as fertile soil, fresh water, sunlight, and petroleum. Some cities, such as New York, may not have large resource base themselves. But these cities provide service to large areas, even globally. It is the amount of resource they can control that determines the level of consumption of particular regions.

With the increasing volatility in commodity prices, more and more people are becoming aware of the importance of natural resources. However, the total cost of gasoline and other commodities is still small for most people. Output from resource industries is still only a small part of overall economic activity, even for major commodity producers, such as Canada. This gives an impression that natural resources constitute only a small part of overall economic activity. But this is really only a matter of definition. The transportation industry is not defined as a resource industry. However the manufacture and operation of vehicles, ships and planes are totally dependent on the availability of energy resources.

To better understand the extent of our dependence on natural resources, we will view our use of natural resources through the lens of thermodynamic theory.

2.2 Natural Resources: Diverse Forms and Unifying Principle

Most of the natural resources on the earth can be attributed to the temperature differential between the sun and the earth. The surface temperature of the sun is 6000 K while the surface temperature of the earth is around 300 K. The high temperature solar surface emits sunlight, which carries high quality energy. The earth receives high quality solar energy and emits low quality waste as infrared light energy. This temperature differential is what drives most things, including living organisms, on the earth. Intuitively, this temperature differential is like the in water levels at a hydro dam, which drives turbines to produce electricity.

Part of solar energy is captured by plants through photosynthesis and converted into chemical energy, which can be stored for a longer period of time than photons. The chemical energy stored in plants can be released to work for plants when and where it is needed to maintain various life activities of the plants, including photosynthesis process. Animals, by eating plants, obtain some of the chemical energy stored in plants. Almost all of the energy sources in the food web on the earth ultimately come from solar energy.

Fresh water is so common that we often take it for granted. In many places, its economic value is very low. However, fresh water is vital for our survival. Fresh water is also very scarce compared with salty water. It comprises only 3 % of all water. The rest, 97 %, is salt water. Salt water has lower free energy level than fresh water. Hence, salt water is stable and in chemical equilibrium, while fresh water is unstable and in a non-equilibrium state. The non-equilibrium state is maintained by solar energy. Solar energy distills salt water into vapor and returns fresh water to the earth, and especially the higher elevation areas, in the form of rain or snow. Since fresh water has higher free energy, the intake of fresh water instead of salt water saves living organisms a lot of energy. That is why areas with abundant fresh water have higher levels of biomass density than areas with little fresh water, such as deserts and oceans, where water is salty. Most cities are located by rivers or lakes, where fresh water is abundant. Fresh water is another gift from the sun.

Water will flow from high to low places due to gravity. Without solar energy enabling water vapor to escape the pull of gravity, all water would remain at low elevations. Clouds would not form and rain would not fall. The whole earth would be a desert. Rivers would not flow. Rivers are vital to human life. Most early civilizations originated by the riverside. Rivers formed the major channels of transportation before the age of cheap fossil fuels. With rivers comes hydro power, which is the only renewable natural resource that generates significant amount of electricity for human society.

Today, most energy needs of human society are provided by fossil fuels. We often call our civilization the fossil fuel civilization. Coal, oil and most of the natural gas are from ancient biological deposits. These biological deposits, which were formed over millions of years, were transformed for use by human societies over the last several hundred years. The abundant use of fossil fuels is the foundation of economic prosperity for many people the world over.

More detailed and systematic discussion about natural resources can be found from the standard references, such as Hall et al. (1986) and Ricklefs (2001). While the forms of natural resources are diverse, most natural resources can be classified into two classes. The first class includes low entropy sources, more popularly understood as energy sources. The second class includes raw materials that can be used as building blocks to harness energy sources for our use. The second class includes most metals. Many resources belong to both classes. Wood can be used as cooking fuels or building material. Petroleum can be processed into vehicle fuels or vehicle parts. Proteins, fats and carbohydrates can be used as energy sources or as building blocks for organisms.

2.3   Living Systems: Positive Return Technologies of Utilizing Resources

Resources are low entropy sources, or materials providing a gradient to the environment. But to utilize resources, it requires structures to harness entropy flows. These structures will be called technologies. We will look at the Periodic Table to see what kinds of chemical elements are good raw materials for building up structures to harness entropy flows.

Table 2.1 The first three row of the periodic table, minus the chemically inactive noble gases

H						
Li	Be	B	C	N	O	F
Na	Mg	Al	Si	P	S	Cl

Table 2.1 shows the first three rows of the Periodic Table without the column for noble gases, which are chemically inactive. Carbon is the first element of the fourth column (the center column of the Periodic Table). Carbon is the lightest of a group of elements that are largely chemically neutral and have four valence bonds. The chemical neutrality and large number of valence bonds makes it easy for carbon to link with many different atoms to form large molecules. Large molecules are essential for the preservation of life. They help to withstand random dissipation (Schrodinger 1944). Large molecules are also essential for performing complex and various tasks. This is why carbon, and carbon alone, is the backbone of life (Atkins 1997). The chemical neutrality of carbon and its four valence electrons makes it easy to combine with and detach from other atoms and to build complex structures. The stable and weak bond of carbon is ideal for storing and releasing energy. This is why carbon is an essential part of any carrier of organic energy, including natural gas and petroleum. Hydrogen is the smallest and the lightest element. That is why hydrogen is a low cost carrier of energy and building block of an organic system.

In order for a technology to last, it must be able to utilize energy resources to make another copy of itself before it wears out. Living systems are technologies that last. Carbon and hydrogen atoms are natural raw materials that make up living systems. Since the buildup of living systems embodies energy resources, living systems themselves become the resources that can be used by other living systems. Animals eat plants, other animals, fungi and bacteria. Bacteria eat plants, animals and other bacteria. Viruses eat bacteria, plants and animals. Human beings and many bacteria also consume fossil fuels, which are transformed from the dead bodies of living systems.

The greatest amount of energy can be accessed by the earth is solar energy. The earliest organisms that developed the structure for photosynthesis were probably bacteria. These bacteria utilized solar energy to build up structures with carbon, hydrogen and other kinds of atoms. Bacteria obtained carbon atoms from carbon dioxide. Initially, bacteria obtained hydrogen from hydrogen sulfide (H_2S) via photosynthesis. Compared with water (H_2O), the chemical bond in hydrogen sulfide is weak. From Table 2.1, both oxygen and sulfur are at the sixth column of the periodic table. Their chemical properties are similar. But sulfur is one row lower than oxygen. It takes less effort, or lower fixed cost, to get hydrogen from hydrogen sulfide. "However, when hydrogen is removed from hydrogen sulfide in the interior of a bacterium, the excrement is sulfur. Sulfur, being a solid, does not waft away, so the colony of organisms has to develop a mode of survival based on a gradually accumulating mound of its own sewage" (Atkins 1997, p. 22). Eventually, organisms developed techniques to break the strong bond in water to get hydrogen.

Oxygen is a byproduct of this technology. Since oxygen is a gas, the pollution does not accumulate locally. It spread out globally. Another advantage of obtaining hydrogen from water is the abundance of water. With water as a raw material in photosynthesis, living systems spread all over the earth.

Oxygen molecules, the waste product from photosynthesis, are highly energetic. This energy destroyed many early living systems and rusts many materials. Eventually, some organisms evolved the structure to harness the energy from oxygen. These organisms used antioxidant technology to reduce the destructive power of the energy from oxygen to tolerable levels. Animals, whose active lifestyle require substantial energy consumption, evolved to take advantage of the abundant atmospheric oxygen. All materials with a gradient against the environment have the capacity to destroy. Only after we develop technologies to harness the gradient of these materials and contain their destructive power to manageable levels, do these materials become resources.

The amount of solar energy available to early photosynthetic bacteria was almost infinite. Those bacteria multiplied quickly. Over time, photosynthetic organisms came to cover most parts of the world. Photosynthesis is probably the most important technology organisms have ever developed. It transforms the vast amount of light energy, which is difficult to store, into chemical energy, which is easy to store. Since the invention of photosynthesis, almost all living organisms have come to derive energy available to them from solar energy, directly or indirectly.

If a technology helps an organism to earn a positive return on resource utilization, the technology will be duplicated and spread out. However, not all technologies developed by organisms are able to provide a non-negative return consistently. Most genetic mutations or innovations are harmful. Even among those mutations where new species are established, over 99.9 % of the species eventually went extinct. Technology and innovation do not guarantee prosperity and safety to any biological species or human societies that adopt them.

A technology is not only expensive to develop but also expensive to maintain. Some fish living in dark underground caves become blind because functional eyes are expensive to maintain. When eyes are no more useful, these structures degenerate. Similarly, the sense of smell of human beings is highly degenerated from that possessed by our ancestors. Walking upright increases our dependence on vision and reduce our dependence on our sense of smell. As a result, smell is under less selection pressure and degenerates. A technology will be developed and maintained if its return on investment is positive during that period. Otherwise, the technology or its host will degenerate.

With the multiplication of organisms, came competition for limited resources. Individual organisms defend themselves or are consumed by other organisms. There are two main strategies, forming social groups to increase the power of the group or enhancing individual power. Social sciences are generally viewed as being exclusively concerned with humanity. But almost all organisms form social groups (Trivers 1985; Willey et al. 2011). Bacteria can live alone. They often thrive better in groups. Most organisms need to develop effective means communicating or coordinating with other organisms. Social biology was developed to study these problems. Studying the social behaviors of other species has provided valuable insights into human behavior. Some single cell organisms evolved into multi-celled organisms. Multi-celled organisms are larger than a single celled organism. An individual multi-cell organism can grow into a collection of many trillion cells. Large animals can often overpower small animals and control more resources. But multi-cell organisms also need to communicate and coordinate among different cells inside their own body, very much like different individuals in a society need to communicate and coordinate with each other. Social and natural sciences need to understand very similar problems.

2.4 Human Technologies and Human Societies

Technologies in human societies are generally understood as tools. These tools exist outside our bodies. The making of tools is not constrained by our body's physical and chemical environment. For example, iron smelting requires a temperature over 1000°, which is much higher than our body temperature. Aluminum smelting requires strong electric currents that would kill us instantly. However, all biological and human technologies are bound by the same economic principle. If a technology generates a positive return in a biological system, it will be developed and maintained. If not, it will decline. Since different technologies require different social structures and bring in different amounts of resources, technical structures and social structures become closely intertwined with each other.

We study past events to estimate the possible patterns in the future and to guide our actions today. The events that greatly affected the lives of people in the past, such as the Industrial Revolution, the Great Depression in 1920s, oil crisis in 1970s and the recent Great Recession, have especially strong grip on the minds of the public. The interpretation of these events greatly influences the public's vision about the future and the policies adopted by governments. In the following, we examine the most general increases over time in humanity's abilities to exploit energy and hence other resources, with a greater efforts to examine more recent events.

2.4.1 The Use of Fire and Cooking

What is the most important activity that separates human beings from other animals? Different people have different answers. From an energy perspective, we might say the use of fire and cooking distinguishes humans from other animals (Wrangham 2009). Cooking kills most of the pathogens in food. Consequently our body doesn't have to spend as much energy on our immune systems. Cooking, by breaking down large organic molecules, predigests food before it is consumed and by breaking down cell walls allows access to much more food. Our body doesn't have to spend as much energy on our digestive systems. The energy saving can be used to nourish other important systems, such as our brain. Human beings can supply more energy to brains for more complex functions. Cooking reduces the risk of infectious disease. It enables human beings to live in higher density, which stimulates the need for better communication. Language, culture and religion flourished to bind people together. Cooking, by using fire regularly, makes us familiar with a chemical process of high energy intensity. This would become crucial in the development of future technologies. It seems no other activity has the same level of impact as cooking and the use of fire to influence the evolution of human beings and human societies.

2.4.2 High Resource Density, Sedentary Lifestyle and Agriculture

Most researchers suggest that sedentary lifestyle followed the advent of agriculture. But it could precede agriculture. Some places have very high concentration of resources. Several times we travelled along Highway 16 toward Prince Rupert, British Columbia. We would pass by a canyon at a place called Morristown. In salmon spawning season, large amount of salmon waited just below the canyon, preparing to jump over the rapids to reach their final spawning destinations. Anyone can use a net to scoop up a fish from the river. The salmon are so abundant that local people are able to live a sedentary life around the canyon year around. They depend on salmon instead of agriculture. Some people could live sedentary life-styles before the development of agriculture. But in general, it provided an incentive to practice agriculture, which requires intensive work in the fields for prolonged periods. So it could be sedentary lifestyle lead to agriculture, instead of the other way around.

While sedentary lifestyle may precede agriculture, large scale sedentary lifestyle followed the invention of agriculture. Agriculture is the growing of a few selected crops with active exclusion of most competing plants. Crops are mainly selected for their nutrition value to humans and not their competitiveness in the fields. They are usually not very competitive. The growth of crops requires intensive human intervention to remove weeds growing in the same fields. Crops, due to their high resource density, become the favored food source for microbes, other animals and other people. Farmers have to remain vigilant against other animals and other people. The increase of resource density always increases the potential of wars against microbes, other animals and other people. Since agriculture produces higher energy yield than hunting and gathering, it can support higher human density. Gradually, farming communities replaced hunting and gathering communities in many places of the world. High human density comes with complex society with many hierarchies.

2.4.3   Bronze Age, Iron Age and the Dark Age

Making bronze tools requires special smelting technology. The advent of the Bronze Age was a significant step in the mastery of using energy resources. The melting point of bronze is 232 °C while the melting point of iron is 1535 °C. It is much more difficult to smelt iron than to smelt bronze. The technology of making iron is developed much later than the technology of making bronze. Iron is much harder than bronze. Weapons made from iron are more powerful than weapons made from bronze. Iron tools can be used to cut down trees, from which charcoal is made. Charcoal is used to smelt iron. More charcoal led to more iron and more iron led to more charcoal. This positive feedback greatly increased the output of energy and iron equipment, such as swords and plows. Swords enabled iron making people to expand their territories. Plows enabled iron making people to get more nutrients from deep soil, thus enhancing crop yields and increasing population density. The arrival of the Iron Age generated the greatest burst of military, economic and cultural activity in human history to that point. However, the amount of iron production was ultimately constrained by the availability of trees around iron mines to provide fuel for the smelting process. A sharp increase in iron production in a place quickly deforested the surrounding area, which limited the scale of iron output throughout most of the Iron Age. Indeed, deforestation and soil erosion often turned once prosperous civilizations into desolate areas. A Dark Age, which consumed fewer resources, followed the Iron Age.

2.4.4   Coal, Iron and the Industrial Revolution

The Industrial Revolution has been interpreted in many ways. Here we offer another interpretation from the interaction between resource and technology, not necessarily inconsistent with other interpretations. Shortly before 1750, a technology of iron making with coal was invented in England. Mining and transportation of coal requires considerable iron made equipment. Since the beginning of the Iron Age, iron was smelted using charcoal, which is made from wood. Before 1750, the need for charcoal in iron making deforested most of England (Jevons 1865). The limit to the supply of charcoal limited the supply of iron, which limited the supply of coal. After the invention of iron smelting by means of coke, derived from coal, it replaced charcoal as the main fuel in iron making. Coal is much more abundant than wood and coal has much higher energy density than wood. More coal led to more iron and more iron led to more coal. The positive feedback between the output of coal and iron, the most important energy source and the most important material of the Iron Age, enabled human beings to grow tremendously in number and in prosperity. This was the essence of the Industrial Revolution. Jevons was very aware of the importance of this invention. In *The Coal Question*, he described it in great detail.

2.4.5 The Age of Oil: The Great Depression, Oil Crises, and the Great Recession

Since the beginning of the Industrial Revolution, the global economy has been growing steadily, most of the time. There have been several slowdowns, such as the Great Depression, the oil crisis in the 1970s and the recent Great Recession. There are many interpretations of the causes of the Great Depression. Very often, it was attributed to structural weaknesses in various economic sectors and policy mistakes by various government agencies. From the energy and technology perspective, the transformation from a coal-based economy, centered on railways, to an oil-based economy, centered on cars and trucks, was partly responsible for the Great Depression. Since hydrogen content of oil is higher than that of coal, oil is a higher quality energy source than coal. In the decade of the 1920s, the number of cars increased tremendously. But the supply of oil had been quite limited. Cars were regarded as a supplement but not replacement to the railroad economy. However around the end of 1920s and the early 1930s, many gigantic oilfields were discovered over a short period of time due to the development of better exploration methods (Deffeyes 2001). It became very clear that the petroleum and car economy would replace the coal and railroad economy. A large part of the railroad economy fell apart immediately upon this realization. But it took time for the car economy to grow enough to replace the railroad economy. The structures of the railway centered economy were very different from that of the highway centered economy. Areas around the train station were often prime real estate and the center of most economic activity in a city in the railway centered economy. In many cities, the street where the train station locates is called the first avenue. However, in a highway centered economy, shopping areas are relocated to malls and residential areas are moved to suburbs. The shift of economic gravity devastated the downtown areas in most cities. Many once prosperous towns built on railways became ghost towns when highways bypassed them.

The discovery of the giant oil fields around 1930 happened over a short period of time. The adjustment was very sudden and painful. The Great Depression was unavoidable, regardless of government policies. Yet the abundance of oil, a higher quality energy source than coal, also set the stage of economic boom after the Second World War, when highways and gas stations were built in many parts of the world. The tremendous growth of the petroleum and car based economy greatly increased the consumption of petroleum. In 1956, M. King Hubbert proposed oil output in US and in the world would eventually peak and decline. Around 1970, many people became concerned by the increasing consumption of resources. A representative work at that time was *The Limits to Growth* published in 1972. In 1970, US oil production peaked. In 1971, the US government stopped converting US dollar to gold at the fixed rate of 35 dollars per ounce, thus delinking the value of dollar to gold, a major commodity. After that, the value of the US dollar depreciated sharply against gold. This put heavy pressure on the price of other commodities, such as petroleum (Galbraith 2008). In 1973, the oil price increased substantially due to the collective action of major oil exporting countries. This increase in oil prices generated a deep recession in many oil importing countries, including most wealthy countries. In 1979, events around Iranian Revolution generated another sharp spike in oil prices, causing another deep and prolonged recession in 1981–1982.

By the 1980s, most wealthy countries that experienced economic recession regained economic growth. The standard explanation is that these countries were able to overcome high oil price with proper economic policies. Hence good economic policy can overcome resource scarcity. However, if we understand human society as a biological system, we will observe that in most wealthy countries, where resource consumption is high, fertility rates dropped below replacement rate after 1973, the year of high oil price. This shows that the biological rate of return has turned negative. However, the drop in the fertility rate temporarily reduced the investment cost of raising the next generation. This generated temporary economic growth for several decades. Eventually, negative biological return will lead to negative monetary return in a society.

Another adjustment for wealthy countries was to move manufacturing activities to poor countries where ordinary people have minimal political power. This greatly reduced the energy consumption in production and in processing waste pollution. This also greatly reduced the salary paid to workers. Lower salary means less consumption of resources. By moving manufacturing to poor countries, the energy consumption is greatly reduced.

In the 1970s, the global oil output was still rapidly growing and the average physical cost of oil production was low. But the control of price setting of oil gradually shifted from major oil consuming countries to oil producing countries. This caused major economic recessions and negative biological return in most wealthy countries. By the end of 1990s, with the rapid depletion of easy to extract oil, the average cost of oil steadily increased. In a now classic paper titled The End of Cheap Oil, Campbell and Laherrere (1998), after carefully examining the oil exploration and production data, concluded "What our society does face, and soon, is the end of the abundant and cheap oil on which all industrial nations depend."

When their paper was published in 1998, oil prices were in the low teens. Since then, the price of oil, as well as other major commodities, has increased substantially. If we had recognized the fundamental importance of resources to the overall economy, we would have stopped the measures to stimulate economy after the burst of the internet bubble in 2000. However, the authorities were convinced that the economic recession in 1973 was caused by improper policy response to high oil prices instead of the high oil price itself. They believed that they had mastered the proper policy response now and were not worried about the steady increase in oil prices. That is why the authorities were unprepared when the financial crisis broke out in 2007 and 2008, although the prices of major commodities had been increasing for some time.

According to neoclassical economic theories, recessions are short term interruptions from long term economic growth. After each recession, economic growth will eventually resume. But from a biological and resource perspective, and from the perspective of heterodox political economy long term economic growth is not assured. It has been for several decades that the biological rate of return has been negative in most wealthy countries. With the demographic structure of inverse pyramid, economic activity will eventually decline. High resource cost makes it more difficult to maintain both high living standard and non negative biological return.

The resource and technology based interpretation provides a simple and consistent interpretation of the major events in human history. There are many other interpretations. Financial crises are often blamed on human greed. This is certainly true. But humans are always greedy, before and after financial crises. Financial crises are often blamed on bad monetary policies, keeping interest rate too low for too long. But in a system with abundant resources, a low interest rate policy only has mild impact to generate inflation. Financial crises are often blamed on wide spread fraud. This is certainly the case in the recent Great Recession. But why is fraud so systematically practiced this time? This is because in an environment with increasing resource cost, fraud becomes the only viable way to generate a high rate of return systematically, still expected by the public who are accustomed to the good old days of abundant resources (Chen and Galbraith 2012b).

2.5 On Inequality

When the water levels inside and outside of a hydro dam are unequal, electricity can be generated. When temperatures inside and outside of an engine are unequal, work can be generated. In popular terms, the inequality in gravitational potential, chemical potential, electric potential and other potentials is called energy. From thermodynamic theory, inequality in potential, or energy is the driving force in the nature. It is also a destructive force. We all need energy provided by oxygen. At the same time, our body produces many antioxidants to prevent oxygen from reacting with and destroying our tissues. Sugar is a vital energy our body needs. But too much sugar in our blood system, and in cells, will damage our health. When we cannot maintain a low level of sugar in our blood system, we get diabetes. Human societies depend very much on high energy input. But we carefully regulate the energy sources. "Playing with fire" is always considered dangerous. Whenever possible, systems with high gradient, or high inequality are carefully regulated. They are contained in isolated or remote places. Furnaces are usually located in basements. Electric generators are usually placed very far from residential areas.

In North America, electric voltage in residential areas is 110 V while in most other parts of the world, the electric voltage is 220 V. To carry the same amount of electric energy in a 110 V system requires much thicker wire than in a 220 V system. But when accidents occur, 110 V causes less shock than 220 V. In a system with abundant natural resources, such as North America, we often choose less resource efficient but safer options. In systems with scarce natural resources, we often choose more resource efficient but riskier options. There is a parallel in social systems. In a social system that controls more resource, its internal inequality is often low. But such system can utilize abundant resources as "energy slaves" (Nikiforuk 2012) or impose inequality on other weaker social systems. In a social system that controls less resource, its internal inequality is often high. In such system, efficiency is very high for the elites, the designers of the system. But such systems also have a higher probability of experiencing violent revolution. When factories are located in wealthy countries, much resource is used to control pollution. This lowers the ratio of output over resource input. But when factories are moved to poor countries, where local population has little political power, little resource is allocated to control pollution. This increases the ratio of output over resource input. Pollution is the reduction of chemical potential. Increasing pollution is the increase of inequality in chemical potential. By increasing inequality, the designers of the system gain higher efficiency and obtain cheaper products as a result.

It requires higher fixed cost to maintain a more unequal society. Dominant parties of a society do not necessarily hope to increase inequality all the time. When the British Empire was expanding rapidly in the 19th century, it abolished slavery, a more extreme form of inequality. By adopting less unequal social systems, Britain was able to maintain and expand a huge empire with relatively little cost and huge profit. The inequality of a system also depends on how long the dominant parties expect the system to last. When we go fishing, we hope to have some inequality over fish. We use a fishing line to hook fish. But if the general public is allowed to use fishing nets in rivers, lakes and oceans, fish population will decline rapidly in a very short period of time. For an unequal system to last, the level of inequality cannot be too extreme. This applies both in nature and in human societies. When the dominant parties expect the system to end soon, the inequality of the social system tends to increase so that dominant parties can extract more profits while the system lasts.

2.6   Carbon and Hydrogen as Energy Sources

Carbon and hydrogen are the main components of organism. They are also the main component of the energy sources in organisms and in fossil fuels. Most energy sources we encounter, from foods we eat to gasoline we use to power our cars, are mainly combinations of carbon and hydrogen atoms. Hydrogen is lighter and has much higher energy density than carbon. Hence hydrogen has lower cost of transportation than carbon. At the same time, hydrogen energy is more costly to produce. For animals, because of their mobile lifestyle, the cost of transportation is high. So animals store a lot of energy as fat, which has a high hydrogen content. Plants, because of their sedentary lifestyle, are more economical to use carbon energy directly. They contain little fat in their bodies, with the exception of seeds, which have high energy demand and need to be more mobile than the plants themselves. This pattern applies to human society as well. Automobiles and airplanes, as transportation tools, are highly mobile. Their energy supplies are petroleum products, such as gasoline, jet fuel and diesel, which contain high hydrogen content and are relatively light. Electricity generators are not mobile. The main energy input in electricity generation is coal, which is mainly carbon, heavy but cheap. The transportation of electric energy is to transport electrons, which are much lighter than atoms. So transporting electricity is a cheap way to transport energy. This is one reason electricity is universally used in most daily activities.

Among main energy resources, natural gas has the highest hydrogen content. That is why natural gas is preferred over coal as the energy source in home for cooking and heating. The last century can be thought as the transformation from carbon economy represented by coal to hydrogen economy represented by oil and natural gas. Therefore, people have moved toward a hydrogen economy long before official mandates from governments. This is also why natural gas and oil, which have much higher hydrogen content than coal, deplete much faster than coal. With the fast depletion of high quality (high hydrogen content) fossil fuel sources, can we create a man-made hydrogen economy?

Compared with coal, oil and natural gas burns more completely and emits fewer pollutants. In general, an energy source that reacts with the environment more easily will leave less harmful residue. At the same time, since clean energy reacts easily in the natural environment, it will be more difficult to preserve. This is why coal is much more abundant than oil and natural gas. In the end, we will have to rely on coal as our main energy supply after the depletion of oil and gas. The twenty-first century will become more a carbon economy instead of a hydrogen economy envisioned in some literature.

When the supply of high hydrogen content energy sources, such as oil and gas, are depleted, a man-made hydrogen energy can be produced from two possible sources. One is from renewable energy, such as solar, wind and biomass. We will discuss this option in a later section. The other is from low quality energy sources, such as coal. From the thermodynamic law, producing certain amount of high quality energy will require more low quality energy. Hence a hydrogen economy will produce more, not less pollution on the global level. The consequence of a hydrogen economy can be seen from an electricity economy. Electricity is a very clean form of energy at end use. Its cleanness enables average households to utilize a huge amount of energy without feeling its negative impact. But power plants are the largest consumer of coal and other energy sources.

A parallel understanding can be made from the separation of residential and industrial areas. The separation makes residential areas cleaner, but adds the extra pollution from transportation as people now need to commute between residential areas and work areas. The longer the distance between residential and industrial or commercial areas, the cleaner the residential areas are. But the total pollution will be higher because the extra pollution caused by transportation will be higher. At the country level, trade allows heavily polluted industries to be moved to poor countries where general population has little political power. While the rich countries enjoy cleaner environment, total pollution on the earth will increase because of the added transportation and communication costs. The concept of ecological footprint, which represents the consumption level of each country, provides a better measurement to the burden of human society to the environment (Wackernagel and Rees 1995). On a global basis our ecological footprint exceeds biocapacity, a situation that cannot be maintained indefinitely.

The prevailing wisdom on energy consumption is that hydrogen based energy should be promoted and the carbon based energy should be suppressed. However, hydrogen based energy is scarce and has already been depleted at fast pace because of their high quality. A further restriction on carbon as an energy source will accelerate the depletion of high quality energy sources and leave future generations in a worse shape. In a more sensible strategy of energy consumption, different energy sources should be utilized in different ways according to the differing physical and chemical properties of hydrogen and carbon. Natural gas, with the highest hydrogen content among carbohydrate fuels, is the cleanest. It can be used as fuels in densely populated residential areas. Gasoline, being a liquid and high energy density fuel because of its high hydrogen content, can be primarily used as transportation fuel, where energy supply has to be carried on vehicles. Coal, being largely carbon, is heavy, abundant and hence cheap. It can be economically used to generate electricity. Utility companies are large companies that can afford to make high fixed cost investment. Since power plants use large amount of fuel, they are in the best position to use expensive equipment to reduce the pollution from coal burning. Before the large price increase of oil during oil crisis in 1970s, many power plants used oil as fuels. But after that most power plants use coal as fuel (Dargay and Gately 2010).

2.7   Some Patterns in Energy Economics

Almost all the energy sources on the earth ultimately come from the sun. However, not all living organisms use solar energy directly. Several factors determine the pattern of energy use. The first important factor is energy density. Solar energy is vast. But the net return from transforming light energy into chemical energy, which organisms can store and transport easily for their further use, is low. Plants, whose sedentary lifestyle requires low level of energy consumption, can effectively utilize solar energy directly through photosynthesis. Most animals, whose mobile lifestyle requires high level of energy input, could not support themselves through photosynthesis internally. Instead, some animals consume plants, which store high density chemical energy transformed from solar energy over a period of time. Other animals eat plant eating animals. Fossil fuels, which are further concentration of biomass in large scale, provide much higher energy density than biomass. The consumption of high energy density fossil fuels is the foundation of economic prosperity enjoyed by human societies in the last several hundred years.

The second factor of energy economics is the relation between the ease of storage and the ease of use. Electricity, which is very easy to use, is very difficult to store. The energy of biomass and fossil fuels are stored in the form of chemical energy, which is less easy to use compared with electricity, is easier to store. Nuclear energy, which requires very expensive system to harness, can be preserved for billions of years. This fundamental tradeoff is determined by the potential well of an energy source. The deeper the potential well, the easier to store the energy and the harder to use it. This is why attempts to reduce the storage cost of some easy to use energy sources, such as electricity, seem so elusive. On the other hand, it is often difficult for easy to store energy, such as fat, to be used easily. That is why it is so difficult for people to lose fat in their body. It is also difficult to achieve high energy density for easy to use energy sources, which react easily due to low potential well. Electricity is easier to use than gasoline. But the energy density in battery, a form of chemical energy, is generally low. For example, the energy density of lead battery is $0.16 * 10^6$ J/kg while the energy density of gasoline is $44* 10^6$ J/kg (Edgerton 1982, p. 74). Great progress has been made to increase energy density of batteries by utilizing smaller atoms, such as lithium. However, potential for further increase of energy density of batteries significantly is limited by the physical properties of electric energy. This is why progresses to develop electric cars that can drive similar distance to gasoline cars without recharging have been slow, although electric cars have a long history.

The third factor of energy economics is the efficiency of energy use and the total consumption of energy. Many people have advocated the increase of efficiency as a way of reduce energy consumption. Will the increase of efficiency reduce overall resource consumption? Jevons made the following observation more than one hundred years ago.

> It is credibly stated, too, that a manufacturer often spends no more in fuel where it is dear than where it is cheap. But persons will commit a great oversight here if they overlook the cost of improved and complicated engine, is higher than that of a simple one. The question is one of capital against current expenditure. ... It is wholly a confusion of ideas to suppose that the economic use of fuel is equivalent to the diminished consumption. The very contrary is the truth. As a rule, new modes of economy will lead to an increase of consumption according to a principle recognized in many parallel instances. (Jevons 1965 (1865), p. xxxv and p. 140)

Put it in another way, the improvement of technology is to achieve lower variable cost at the expense of higher fixed cost. Since it takes larger output for higher fixed cost systems to breakeven, to earn a positive return for higher fixed cost systems, the total use of energy has to be higher than before. That is, technology advancement in energy efficiency will increase the total energy consumption. Jevons' statement has stood the test of time. Indeed, the total consumption of energy has kept growing, almost uninterrupted decades after decades, in the last several centuries, along with the continuous efficiency gain of the energy conversion (Inhaber 1997; Smil 2003; Hall 2004).

Figure 2.1 displays the total primary energy consumption worldwide from 1965 to 2012, a period of rapid technology progress. During this period, energy consumption grew steadily, with only two brief interruptions. From 1979 to 1982, a period of Iranian Revolution, oil price jumped from 13.60 US dollars per barrel in 1978 to 35.69 in 1980, causing serious recession in industrial world. The drop of

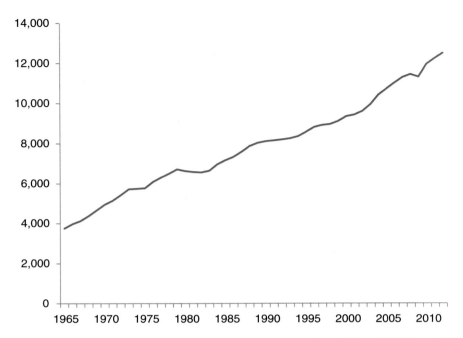

Fig. 2.1 Global energy consumption (Million ton oil equivalent) from 1965 to 2012. *Source* BP

energy consumption from 1979 to 1982 was due to sharp jump in oil price and the ensuing contraction of economic activities, not due to technology progress. Another brief period of interruption is year 2009. In 2008, oil price rose sharply, reaching 147 dollars per barrel at its peak. The ensuing Great Recession diminished the capacity for energy consumption. But the decline of energy consumption only last for one year.

We may also examine energy consumption of individual innovations such as hybrid cars. Hybrid cars need two engine systems, such as an internal combustion engine and electronic motor. Hence their production requires more resource and energy input. If the owners of hybrid cars drive very little, the total energy cost of a hybrid car is actually higher than a conventional car due to the high resource consumption in manufacturing hybrid cars. Only when the owners of hybrid cars drive extensively, hybrid cars become more economical than conventional cars. Hence the use of more efficient cars will encourage, indeed enforce, the high consumption of energy. It should also be noted that hybrid cars, equipped with two engines, are heavier than conventional cars of the same class. Hence hybrid cars will be less fuel efficient in highway driving where frequent stop and acceleration is not required. In the end, hybrid cars, because of their high resource consumption in production and hence high prices, will become a symbol of status, just like SUV at present time.

2.8   On the Concept of Renewable Energy

The deepening public concern about the long-term availability of fossil fuels has promoted governments worldwide to adopt policies to subsidize research and production of renewable energy. To understand the effectiveness of these policies, we need to clarify the concept of renewable energy.

Almost all life forms depend on solar energy directly or indirectly for billions of years. So the use of renewable energy is not a new adventure. Instead, it has been practiced for billions of years by all life forms.

Almost all of the energy sources are renewable to some degree. Fossil fuels are generally considered to be non-renewable because of the time, heat, and pressure needed to transform carbohydrates into hydrocarbons. They are produced every day in various geological structures, however the rate of production of fossil fuels is much lower than the rate of consumption. A resource is renewable when its consumption rate is lower or equal to its regeneration rate. Otherwise it is not renewable. Hence, the concept of renewable resource is intrinsically linked to the level of consumption of that resource. For example, many supposed renewable resources, such as fishery, collapsed when consumption levels became higher than regeneration. This will clarify a lot of confusions around renewable resources.

Corn is supposedly an important part of renewable energy industry. Pimentel and Patzek (2005) reviewed the past research works on the production of ethanol from corn and found the following results.

1. The total energy input to produce a liter of ethanol is higher than the energy value of a liter of ethanol. Thus there is a net energy loss in ethanol production from corn.
2. Producing ethanol from corn causes major air, water pollution and degraded environmental system.
3. There are over 3 billion people in the world are malnourished. Expanding ethanol production, which divert corn needed to feed people, raises serious ethical issues.

Despite the fact that ethanol production from corn is not a renewable energy, it pollutes environment and raise serious ethical issues, government subsidy on bio-energy has been expanding rapidly over time. Why is that? There could be several reasons.

1. Because of government subsidy, companies participating bio-energy production have been very profitable (Pimentel and Patzek 2005). Hence they will lobby hard for this type of subsidy.
2. The term "renewable resources" provides a psychological satisfaction for the general public. People can continue their grand lifestyle that requires high level of resource consumption while still feel morally superior because they invest on renewable resources, which are supposedly to be beneficial to future generations.

3. Governments are eager to be seen as doing "something" for the environment. Spending money on environment related issues projects a positive image. Government actions are generally channeled into directions with maximum political support and minimum political resistance. This means that government policies often benefit the current generation at the expense of future generations, who are not here to vote and lobby. This is true even for government policies that are supposed to benefit future generations.

Like any other viable investment in life, viable investment in energy should yield positive rate of return. Estimates of energy return on investment (EROI) of many renewable energy resources have been produced by experts in different areas. But the origin of the idea, and most of the numbers, are attributable to Charles Hall. Sometimes these estimates are very high. For example, many estimations of energy return on investment of wind power is around 20 and some estimations of energy return on investment of solar panel is around 7. If the energy returns of these resources are really that high, the construction of these projects will spread like wild fire, even without any government subsidy. There will be no more problem of energy shortage because of the vast amount of solar energy. It is beyond my expertise to estimate the rate of return on each type of renewable energy resources. For more background information, please see Prieto and Hall (2013), Hall et al. (2014), Lambert et al. (2014). Instead, I will discuss some general economic principles that suggest the costs of most renewable energy programs have been seriously underestimated in some literature.

Living organisms, including human beings, generally utilize resources from easy to difficult, or in economic terms, from low fixed cost to high fixed cost, or from most accessible to least accessible For example, humans use wood, coal, oil and natural gas, in that order, because the fixed cost of using them increases in each case. Promising renewable energy sources have not been brought into market place because of their high fixed costs. These high fixed costs are heavily subsidized through tax dollar funded university and industry research. The development and maintenance of most renewable energy technology requires high level of technology expertise, which is developed through the expensive education system. While education is funded by general tax revenue, it mainly benefits high tech industries.

All proponents of alternative renewable energy acknowledge the high fixed cost inherent in these new energy resources and argue that scale economy will eventually bring down the average cost. There are several types of economies of scale in the resource industry. We will discuss them separately.

The first type of scale economy is in high-tech research. The expensive high tech research only pays off when it market size is large. For example, people have been harnessing wind power for many centuries. But only with light and strong new materials such as fiberglass, has generating electricity via wind power become feasible. Fiberglass, as a synthetic material made from petroleum, is a direct outgrowth from the fossil fuel industry. Indeed, most of metal based technologies are supported by the abundance of fossil fuels. When the fossil fuel becomes scarce, most modern industries, with their high fixed cost, may not have a sufficiently large scale to be viable. While the fixed costs in developing some renewable energy seem high today, they could be much higher in the future when fossil fuels become less abundant.

The second type of economy of scale is the use of large quantities of fossil fuel itself. For example, electrical transmission systems are very expensive to build and maintain. They become economically viable only when large amount of electricity, most of which is generated by fossil fuels, is being transmitted. It is the scale economy of fossil fuel generated electricity that supports the infrastructure that is also been used by alternative energy sources. If fossil fuels are excluded in electricity generation or depleted in the future, will the alternative energy sources be able to provide sufficient large market size to support the high cost electrical transmission system? This leads to the third type of scale economy: the scale of each type of renewable energy.

Hydro power, which generates 7 % of total electricity output worldwide, has already achieved substantial economy of scale. At certain locations, hydro power has the advantage of high energy density, just like fossil fuels. This is also why hydro power has achieved such a large scale. For other renewable energy resources, such as biomass, solar and wind, the energy density may not be very high and steady, which poses a physical limit on the reduction of cost.

The current biosphere is the result of more than three billion years of evolution. In the competition for survival, many different ways of utilizing resources more economically have been explored. The most discussed forms of renewable energy, such as solar, wind, and bio energy, have been around for billions of years and have been extensively explored by many species, including human beings. While it is difficult to rule out the possibility that human beings can develop new technologies that has significantly higher overall efficiency in energy use than other living organisms and our own ancestors, if research and development costs are included, the likelihood will be low. Human beings, like other dominant species, excel at controlling more resources, not at utilizing resources more efficiently (Colinvaux 1980). Furthermore, research activities themselves are very resource intensive and accelerate the depletion of natural resources. Hence government policies about future energy investment patterns should not rest on the assumption that technological progress will automatically substitute the demand for natural resources, as mainstream economic theory asserts (Samuelson and Nordhaus 1998, p. 328).

2.9   Concluding Remarks

Because of the importance of energy in our life, people have pursued the dream of convenient and cheap renewable energy since time immemorial. In the course of history, many people believe they have discovered an inexhaustible source of energy, such as battery, or invented one or another kind of perpetual motion machine. They think their discoveries or inventions can be put into practical use in large scale once necessary technical improvements can be made in the future.

However, more rigorous investigation leads to the development of thermodynamic theory, which rules out the possibility of a perpetual, or renewable energy source without external input.

In this chapter, I investigated further how physical environment enables and constrains living organisms and economic systems by integrating the economy of human society into the economy of nature. I explored the relation between natural resources and technology in human society. It helps us envision the future of human society in an environment of increasingly scarce and costly natural resources. The main results can be summarized as follows. First, the survival and prosperity of human society depends entirely on the availability of natural resources. Second, while the forms of natural resources are diverse, they can be understood from the unifying principle as low entropy sources. Third, to utilize natural resources, fixed structures are required, which consume resources themselves. Fourth, when certain structures can generate positive returns on the use of natural resources over an indefinite period of time, these structures are called living organisms. Fifth, it is the unique chemical properties of carbon that enables it to become the backbone of life. The major non-renewable resources that our industrial civilization builds on, such as coal, petroleum and natural gases, are generated from the remains of the living organisms. They all contain carbon.

Some practical implications emerge from our theoretical discussion. First, we prefer high quality resources over low quality resources. This is why we moved gradually from more carbon based fuels, such as coal, to less carbon based fuels, such as natural gas and petroleum. However, as high quality resources, such as conventional oil, are seriously depleted, human society will be forced to move back toward a carbon based economy from the current mixed carbon and hydrogen economy. This contradicts the often dreamed of hydrogen economy in the future. Second, increasing energy efficiency, which requires the increase of fixed cost, will increase total resource consumption. This was pointed out by Jevons more than a hundred years ago. Third, due to the levels of potential well, energy sources that are easy to use, such as electricity, are difficult to store. This is why it is so difficult to develop electric cars that can drive long distance without recharging.

Chapter 3
Production: A Mathematical Theory

3.1 A Historic Review of Related Ideas and Mathematical Techniques

Because of the fundamental link between thermodynamics and life, many attempts have been made to develop analytical theories based on the principle of thermodynamics and apply them to living systems and human society. Two of the influential theories are Lorenz' chaos theory and Prigogine's far from equilibrium thermodynamic theory. Lorenz, a meteorologist, simplified weather equations, which are thermodynamic equations, into ordinary differential equations. He found chaos properties from these equations. Prigogine developed the theory from some chemical reactions. Ping Chen has written extensively to apply Prigogine's theory to social sciences, offering great insights into many social problems (Chen 2010). The theories of Lorenz and Prigogine greatly influenced the thinking in biology and social sciences. However, these theories, as well as other related mathematical theories, do not model life process or social activities directly.

Since uncertainty is an integral part of life processes, the advancement of stochastic calculus is essential for the development of an analytical thermodynamic theory of life and human society. In the past several decades, some fundamental works in the area of stochastic calculus were undertaken by people with very diverse backgrounds. Three works are particularly relevant to the development of our theory. The first is Ito's Lemma, which provides a rule to find the differential of a function of stochastic variable. Ito's Lemma was obtained in 1940s. But its importance was not recognized until its wide spread application in financial economics several decades later.

The second tool is Feynman-Kac formula, which maps a stochastic process into a deterministic thermodynamic equation. In natural science, there is a long tradition of studying stochastic processes with deterministic partial differential equations. For example, heat is a random movement of molecules. Yet the heat process is often studied by using heat equations, a type of partial differential equations. Richard

© Springer Science+Business Media New York 2016
J. Chen, *The Unity of Science and Economics*,
DOI 10.1007/978-1-4939-3466-9_3

Feynman (1948) attempted to simplify calculation in quantum mechanics by transforming problems in stochastic processes into problems in deterministic processes. The new mathematical technique enabled him to perform many computations in quantum mechanics which were very difficult in the past. With this he established the theory of quantum electrodynamics. The breakthrough in physics is often generated by the breakthrough in new mathematical methods, which enables us to describe the subtler parts of the nature. An important motivation in Feynman's research was his seek for universality. "The question that then arose was what Dirac had meant by the phrase 'analogous to,' and Feynman determined to find out whether or not it would be possible to substitute the phrase 'equal to.'" (Feynman and Hibbs 1965, p. viii). Kac (1951) extended Feynman's method into a mapping between stochastic process and partial differential equations, which was later known as the Feynman-Kac formula. Despite its highly technical nature, Feynman-Kac formula is a very general result and has proved to be extremely useful in many different fields (Kac 1985).

Kolmogorov developed more systematic approach to stochastic calculus (Kolmogorov 1931). He defined two different types of mappings from stochastic processes to deterministic differential equations as backward equations and forward equations. Kolmogorov backward equations are equivalent to Feynman-Kac formula. Kolmogorov forward equations are equivalent to Fokker-Planck equation in physics. Two types of equations have different physical meanings. It will be very important to distinguish them in specific applications, as we will see later.

The third is Black-Scholes (1973) option pricing theory, which provides an analytical formula of observable variables to price a financial instrument whose payoff depends on a stochastic process. This is a landmark contribution in social sciences. It shows that a complex economic problem can be effectively modeled by a simple analytical theory and much information about it can be obtained through such an analytical theory. Fischer Black, one of the co-developers of the Black-Scholes theory, was a legendary figure in finance. Jack Treynor, who introduced Fischer Black to the field of finance, had the following observation:

> Fischer never took a course in either economics or finance, so he never learned the way you were supposed to do things. But that lack of training proved to be an advantage … since the traditional methods in those fields were better at producing academic careers than new knowledge. Fischer's intellectual formation was instead in physics and mathematics, and his success in finance came from applying the methods of astrophysics. Lacking the ability to run controlled experiments on the stars, the astrophysist relies on careful observation and then imagination to find the simplicity underlying apparent complexity. In Fischer's hands, the same habits of research turned out to be effective for producing new knowledge in finance. (Mehrling 2005, p. 6)

Jack Treynor (1996) summarized:

> Fischer's research was about developing … insightful, elegant models that changed the way we look at the world. They have more in common with the models of physics — Newton's laws of motion, or Maxwell's equations — than with the econometric "models" — lists of loosely plausible explanatory variables — that now dominate the finance journals.

In the Black-Scholes theory, the movement of price of financial assets, S, is modeled with lognormal processes

$$\frac{dS}{S} = rdt + \sigma dz .$$

where r is the rate of expected return and σ is the rate of uncertainty. The price of a financial derivative, C, as a function of prices of its underlying assets, satisfy the following Black-Scholes equation.

$$\frac{\partial C}{\partial t} + rS\frac{\partial C}{\partial S} + \frac{1}{2}\sigma^2 S^2 \frac{\partial^2 C}{\partial S^2} = rC$$

I had been thinking about an analytical thermodynamic theory of life systems for many years when I learned about the Black-Scholes theory. The most fundamental property of life is their ability to extract low entropy from the environment to compensate continuous dissipation. Soon I realized this property can be represented by lognormal processes, where r is the rate of extraction of low entropy and σ is the rate of diffusion. Every stochastic process can be mapped into a deterministic thermodynamic equation, which is often easier to handle and yields more results. So I hope Black-Scholes equation and option theory may offer some insight for an analytical thermodynamic theory of life systems. After several years, I first developed such a theory based an analogy between option theory and living systems. Later I was able to derive the theory directly without depending on its analogy with option theory. In the next section, we will provide an updated version of this theory.

3.2 A Mathematical Theory of Production

The theory described in this section can be applied to both biological and economic systems. For simplicity of exposition, we will use the language of economics. However the extension to biological system is straight forward.

We start the investigation by asking: What are the most fundamental properties of organisms and organizations? How do we represent these fundamental properties in a mathematical theory? First, organisms and organizations need to obtain resources from the environment to compensate for the continuous diffusion of resources required to maintain various functions. This fundamental property can be represented mathematically by lognormal processes, which contain both a growth term and a dissipation term.

Suppose S represents the amount of resources accumulated by an organism or the unit price of a commodity, r, the rate of resource extraction or the expected rate of change of price and σ, the rate of diffusion of resources or the rate of volatility of price change. Then the process of S can be represented by the lognormal process

$$\frac{dS}{S} = rdt + \sigma dz.$$ (3.1)

where

$dz = \varepsilon\sqrt{dt}$, $\varepsilon \in N(0, 1)$ is a random variable with standard Gaussian distribution.

   The process (1) is a stochastic process. Although a stochastic process will generate many different outcomes over time, we are mostly interested in the average outcomes from such processes. For example, although the movement of individual gas molecules is very volatile, air in a room, which consists of many gas molecules, generates a stable pressure and temperature. We usually study the average outcomes of stochastic processes by looking at the averages of the underlying stochastic variables and their functions. These investigations often transform stochastic processes into their corresponding deterministic equations. For example, heat is a random movement of molecules. Yet the heat process is often studied by using heat equations, a type of deterministic partial differential equation.

   Feynman (1948) developed a method of averaging stochastic processes under very general conditions, which is usually called path integral. Kac (1951) extended Feynman's method into a mapping between stochastic processes and partial differential equations, which was later known as the Feynman-Kac formula. According to the Feynman-Kac formula (Øksendal 1998, p. 135), if

$$C(t, S) = e^{-qt}E(f(S_t))$$ (3.2)

is the expected value of a function of S at time t discounted at the rate q, then $C(t, S)$ satisfies the following equation

$$\frac{\partial C}{\partial t} = rS\frac{\partial C}{\partial S} + \frac{1}{2}\sigma^2 S^2 \frac{\partial^2 C}{\partial S^2} - qC$$ (3.3)

with

$$C(0, S) = f(S)$$ (3.4)

   It should be noted that many functions of S satisfy Eq. (3.3). The specific property of a particular function is determined by the initial condition (3.4). This is similar to the Black-Scholes option theory. The Black-Scholes equation is satisfied by any derivative securities. It is the end condition at contract maturity that determines the specific property of a particular derivative security.

   Second, for an organism or an organization to be viable, the total cost of extracting resources has to be less than the gain from the amount of resources extracted, or the total cost of operation has to be less than the total revenue. Costs include fixed cost and variable cost. In general, production factors that last for a long time, such as capital equipment, are considered fixed cost while production factors that last for a short time, such as raw materials, are considered variable costs. If employees are on long term contracts, they may be better understood as fixed costs, although in the economic literature, they are usually classified as variable costs. Typically, a lower variable cost system requires a larger investment in fixed costs, though the converse is not necessarily true. Organisms and organizations can adjust their level of fixed and variable costs to achieve high level of return on their investment. Intuitively, in a large and stable market, firms will invest heavily in fixed cost to reduce variable cost, thus achieving a higher level of economy of scale. In a small or volatile market, firms will invest less in fixed cost to maintain a high level of flexibility. In the following, we will examine the relation between fixed cost and variable cost in a very simple project.

Suppose there is a project with a duration that is infinitesimally small. It only has enough time to produce one unit of product. If the fixed cost is lower than the value of the product, in order to avoid arbitrage opportunity, the variable cost should be the difference between the value of the product and the fixed cost. If the fixed cost is higher than the value of this product, there should be no extra variable cost needed for the product. Mathematically, the relation between fixed cost, variable cost and the value of product in this case is the following:

$$C = \max(S - K, 0) \tag{3.5}$$

where S is the value of the product, C is the variable cost and K is the fixed cost of the project. When the duration of a project is of a finite value T, relation (3.5) can be extended into

$$C(0, S) = \max(S - K, 0) \tag{3.6}$$

as the initial condition for Eq. (3.3). Equation (3.3) with initial condition (3.6) can be solved to obtain

$$C = Se^{(r-q)T}N(d_1) - Ke^{-qT}N(d_2) \tag{3.7}$$

where

$$d_1 = \frac{\ln(S/K) + (r + \sigma^2/2)T}{\sigma\sqrt{T}}$$

$$d_2 = \frac{\ln(S/K) + (r - \sigma^2/2)T}{\sigma\sqrt{T}} = d_1 - \sigma\sqrt{T}$$

The function $N(x)$ is the cumulative probability distribution function for a standardized normal random variable. From (3.6), the solution of the Eq. (3.3) can be interpreted as the variable cost of the project. However, we will investigate shortly whether the function represented in formula (3.7) has common properties of variable costs. For a given investment problem, different parties may select different discount rates. To simplify our investigation, we will make the discount rate equal to the expected rate of growth. This is to set

$$q = r \tag{3.8}$$

This choice of discount rate can be understood from two perspectives. From a biological perspective, fast growing organisms also have a high probability of death. In a steady state, the growth rate has to be equal to the death rate. In the biological literature, the discount rate is usually set equal to the growth rate (Stearns 1992). From the perspective of economics, in option theory, the discount rate is set equal to the risk free rate. The level of risk of an option contract is represented by implied volatility, which does not necessarily equate with past volatility or future expected volatility. Some people do not agree with the economic logic behind the mathematical derivation of Black-Scholes equation that made the risk related discount rate disappear (Treynor 1996). However, the disappearance of the separate discount rate greatly simplified our understanding of how option values are related to market variables. From both a biological and economic perspective, this choice of discount rate provides a good starting point for further investigation.

With q equals to r, Eq. (3.3) becomes

$$\frac{\partial C}{\partial t} = rS \frac{\partial C}{\partial S} + \frac{1}{2}\sigma^2 S^2 \frac{\partial^2 C}{\partial S^2} - rC \tag{3.9}$$

and solution (3.7) becomes

$$C = SN(d_1) - Ke^{-rT}N(d_2) \tag{3.10}$$

Formula (3.10) provides an analytical formula of C, variable cost as a function of S, product value, K, fixed cost, σ, uncertainty, T, duration of project and r, discount rate of a firm. Similar to understanding in physics, the calculated variable cost is the average expected cost of variable inputs. With an analytical formula, we can calculate how variable cost changes with respect to other major factors in economic and biological activities. Formula (3.10) takes the same form as the Black-Scholes formula for European call options. But the meanings of the parameters in this theory differ from that in the option theory.

We will briefly examine the properties of formula (3.10) as a representation of variable cost. First, the variable cost is always less than the value of the product when the fixed cost is positive. No one will invest in a project if the expected variable cost is higher than the product value. Second, when the fixed cost is zero, the expected variable cost is equal to the value of the product. When the fixed cost approaches zero, the expected variable cost will approach the value of the product. This means that businesses need to make a fixed investment before they can expect a profit. Similarly, all organisms need to invest in a fixed structure before they can extract resources profitably. Some do not agree with this statement and provide examples of low fixed cost investment with high profits, such as J. K. Rowling writing Harry Potter books. Our results are about the statistical average. While a small percentage of authors earn high incomes from blockbusters, an average author does not earn a high income. Third, when fixed costs, K, are higher, variable costs, C, are lower. Fourth, for the same amount of the fixed cost, when the duration of a project, T, is longer, the variable cost is higher. This shows that investment value depreciates with time. Fifth, when risk, σ, increases, the variable cost increases. Sixth, when the discount rate becomes lower, the variable cost decreases. This is due to the lower cost of borrowing. All these properties are consistent with our intuitive understanding of and empirical patterns in production processes.

After obtaining the formula for the variable cost in production, we can calculate the expected profit and rate of return of an investment. Suppose the volume of output during the project life is Q, which is bound by production capacity or market size. During the project life, we assume the present value of the product to be S and the variable cost to be C. Then the total present value of the product and the total cost of production are

$$SQ \text{ and } CQ + K \tag{3.11}$$

respectively. The net present value of the project is

$$QS - (QC + K) = Q(S - C) - K \tag{3.12}$$

The rate of return of this project can be represented by

$$\frac{QS - (K + QC)}{K + QC} = \frac{QS}{K + QC} - 1 \tag{3.13}$$

It is often convenient to represent S as the value of output from a project over one unit of time. If the project lasts for T units of time, the net present value of the project is

$$TS - (TC + K) = T(S - C) - K \tag{3.14}$$

The rate of return of this project can be represented by

$$\frac{TS - (K + TC)}{K + TC} = \frac{TS}{K + TC} - 1 \tag{3.15}$$

Unlike a conceptual framework, this mathematical theory enables us to make quantitative calculations of returns of different projects under different kinds of environments. Jack Treynor's comment about Black-Scholes theory provides a relevant background to understand our production theory (Treynor 1996):

Time has always been a pesky problem for economists, who have dealt with it by

1. Restricting their model to perpetuities (Modigliani and Miller).
2. Focusing on one-period problems (Markowitz's portfolio balancing model).

3. Reducing the dynamic flow of economic events to a static long run and a static short run (Alfred Marshall).

That these pioneers in quantifying the previous unquantifiable ducked the problems is a measure of what Black-Scholes accomplished.

Our production theory, as an extension of the Black-Scholes methodology, can be applied directly to refine the theories of these pioneers. By considering corporate finance problems in a finite time horizon, we are able to provide a more precise understanding of the problems related to capital structure considered by Modigliani and Miller (Chen 2006a). By working in a continuous time framework instead of a one period framework, we are able to obtain a more refined understanding of relations among risk, discounting and duration of projects. This helps us understand patterns such as hyperbolic discounting (Ainslie 1992). By identifying long run cost as fixed cost and short run cost as variable cost and establishing their relations, we provide an analytical theory of economic dynamics that was conceived by Alfred Marshall qualitatively.

Soon after Black-Scholes (1973), it became apparent to many researchers that similar approaches may be applied to capital investment. These approaches are generally called the real option theory. The book by Dixit and Pindyck (1994) is the acknowledged classic in real option theory. In that book, many partial differential equations were derived, but no analytical results about the key factors in capital investment were obtained. As a result, the real option theory "either use stylized numerical examples or adopt a purely conceptual approach to describing how option pricing can be used in capital budgeting" (Megginson 1997, p. 292). In comparison, the production theory presented here provides simple analytical formulas for the key parameters in capital investment. A detailed literature review and comparison between this theory and the real option theory was provided in Chen (2006b).

From a mathematical perspective, Dixit and Pindyck (1994) adopted Kolmogorov forward equations in their book. This is a little bit surprising, for in their book, they stated, "Feynman could be claimed as the father of financial economics" (Dixit and Pindyck 1994, p. 123). Probably they were not aware that Kolmogorov forward equations and Feynman's method were different methods. Currently, Kolmogorov forward equations, which are called Fokker–Planck equations in physics, are more widely used in economic theories (Aoki and Yoshikawa 2006). Kolmogorov forward equations describe the evolution of probability distributions of a system while Kolmogorov backward equations describe the evolution of average values of a system. Since many decision makings are about the average values of gains or losses, Kolmogorov backward equations and Feynman-Kac formula provide natural representations in many important economic problems.

Chen (2010) questioned the validity of using lognormal processes in modeling economic processes. He suggested that lognormal processes grow exponentially while no economic system will grow forever. He is certainly right that no economic system grows forever. But our theory is concerned about biological and economic processes over each generation or duration of a project, which lasts for a finite period of time. Over a finite period of time, biological and economical systems can grow exponentially. So our theory is not inconsistent with basic scientific principles.

Since our theory is very similar to Black-Scholes option theory, it is essential to compare the basic equations of Black-Scholes theory and our theory. The Black-Scholes equation is

$$-\frac{\partial C}{\partial t} = rS\frac{\partial C}{\partial S} + \frac{1}{2}\sigma^2 S^2 \frac{\partial^2 C}{\partial S^2} - rC \qquad (3.16)$$

The basic equation in our theory is

$$\frac{\partial C}{\partial t} = rS\frac{\partial C}{\partial S} + \frac{1}{2}\sigma^2 S^2 \frac{\partial^2 C}{\partial S^2} - rC \qquad (3.9)$$

The Black-Scholes equation has a negative sign in front of the time derivative. From the physics perspective, our equation represents a thermodynamic process while the Black-Scholes equation represents an inverse thermodynamic process. From the economic perspective, Black-Scholes equation solves the current price of derivative securities when the future payout is determined while our equation solves the expected variable cost in the future when the current fixed cost is determined. These two theories solve different problems in economic activities.

This theory provides quantitative relations about major factors in economic and biological systems. First we will examine how variable costs change with fixed cost at different levels of uncertainty. By calculating variable costs from (3.10), we find that, as fixed costs are increased, variable costs decrease rapidly in a low uncertainty environment and change very little in a high uncertainty environment. To put it in another way, high fixed cost systems are very sensitive to the change of uncertainty level while low fixed cost systems are not. This is illustrated in Fig. 3.1.

The above calculation indicates that systems with higher fixed investment are more effective in a low uncertainty environment and systems with lower fixed investment are more flexible in high uncertainty environment. Mature industries, such as household supplies, are dominated by established large companies such as Proctor &Gamble while innovative industries, such as IT, are pioneered by small and new firms. Microsoft, Apple, Yahoo, Google, Facebook and countless other innovative businesses are started by one or two individuals and not by established firms. Similarly, in scientific research, mature areas are generally dominated by top researchers from elite schools, while scientific revolutions are often initiated by newcomers or outsiders (Kuhn 1996).

We now discuss the returns of investment on projects of different fixed costs with respect to the volume of output or market size. Figure 3.2 is the graphic representation of (3.13) for different levels of fixed costs. In general, higher fixed cost projects need higher output volume to breakeven. At the same time, higher fixed cost projects, which have lower variable costs in production, earn higher rates of return in large markets.

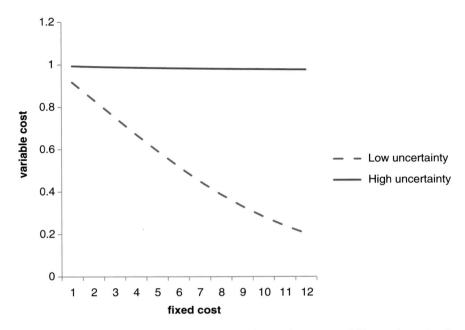

Fig. 3.1 Fixed cost and uncertainty: In a low uncertainty environment, variable cost drops sharply as fixed costs are increased. In a high uncertainty environment, variable costs change little with the level of fixed cost

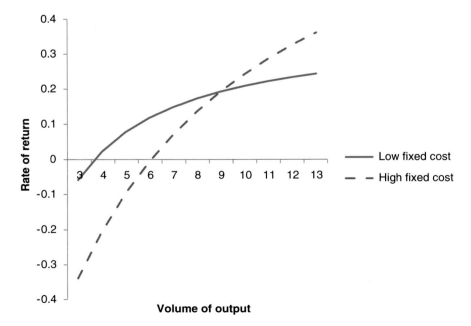

Fig. 3.2 Fixed cost and the volume of output: For a high fixed cost investment, the breakeven market size is higher and the return curve is steeper. The opposite is true for a low fixed cost investment

We can see from the above discussion that the proper level of fixed investment in a project depends on the expectation of the level of uncertainty and the size of the market. When the outlook is stable and the market size is large, projects with high fixed investment earn higher rates of return. When the outlook is uncertain or market size is small, projects with low fixed cost break even easier, or at least more quickly.

In the ecological system, the market size can be understood as the size of resource base. When resources are abundant, an ecological system can support large, complex organisms (Colinvaux 1978). Physicists and biologists are often puzzled by the apparent tendency for biological systems to form complex structures, which seems to contradict the second law of thermodynamics (Schneider and Sagan 2005; Rubí 2008). However, once we realize that systems of higher fixed cost provide higher return in the resource rich and stable environments, this evolutionary pattern becomes easy to understand. An example from physiology will highlight the tradeoff between fixed and variable cost with different levels of output.

> An increased oxygen capacity of the blood, caused by the presence of a respiratory pigment, reduces the volume of blood that must be pumped to supply oxygen to the tissues. ... The higher the oxygen capacity of the blood, the less volume needs to be pumped. There is a trade-off here between the cost of providing the respiratory pigment and the cost of pumping, and the question is, Which strategy pays best? It seems that for highly active animals a high oxygen capacity is most important; for slow and sluggish animals it may be more economical to avoid a heavy investment in the synthesis of high concentrations of a respiratory pigment. (Schmidt-Nielson 1997, p. 120)

For high output systems (highly active animals) investment is fixed cost (respiratory pigment) is favored while for low output systems (slow and sluggish animals) high variable cost (more pumping) is preferred. Pumping is variable cost compared with respiratory pigment because respiratory pigment lasts much longer.

In the next section, we will provide some examples to apply the mathematical theory to understand different economic phenomena. In Sect. 3.4, we will apply the theory to understand monetary policies and business cycles, which are among the most important problems in economic theories.

3.3 Several Numerical Examples

With a simple analytical theory, we can easily observe the impacts of particular parameters or relations among parameters from simple calculations. In the following, we will present some examples.

All parameters in our theory, except uncertainty, correspond to directly observable quantities. This is very similar to option pricing theory, in which all parameters, except volatility, correspond to directly observable quantities. In option pricing theory, volatility is often called implied volatility because volatility is implied from the option prices. Similarly in our theory, uncertainty is implied from the expected variable costs. Indeed, the value of uncertainty rate can involve many factors. The meaning of uncertainty rate can be very different in different applications.

3.3.1   Economy of Scale and the Law of Diminishing Return

All economic systems experience economy of scale and the law of diminishing return at the same time. This can be modelled by setting uncertainty as a function of output. This can be modeled with uncertainty, σ, as an increasing function of the volume of the output. Specifically, we can assume

$$\sigma = \sigma_0 + lQ$$

where σ_0 is the base level of uncertainty, Q is the volume of output and $l > 0$ is a coefficient. Intuitively, when the size of a company increases and the business expands, the internal coordination and external marketing becomes more complex. With the new assumption, we can calculate the rate of return of production from formula (3.13). The result from the calculation is presented in Fig. 3.3. From Fig. 3.3, the rate of return initially increases with the production scale, which is well known as the economy of scale. When the size of the output increases further, the rate of return begin to decline. This is the law of diminishing return. In specific applications, we can analyze how each factor influence the shape of return curves and try to obtain high rate of returns for our investment.

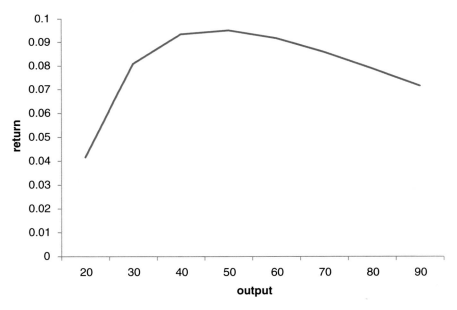

Fig. 3.3 Volume of output and the rate of return: The rate of return of a project with respect to volume of output, when uncertainty is an increasing function of volume of output

3.3.2   Increasing Fixed Cost to Reduce Uncertainty

When the fixed cost of a system increases, the increased fixed cost can often help reduce uncertainty. Many organisms, from stegosaurus to turtles, invest in armor to decrease predation. Air conditioning and heating systems can reduce the temperature uncertainty of a building. But the expansion of air conditioning requires an increase in electricity consumption. Insurance can reduce uncertainty of large losses for the individual policy holder. Yet, as the Great Financial Crisis of 2008–2009 showed, insurance companies themselves (for example AIG) can teeter on the edge of bankruptcy and require bailouts when faced with systemic failure. This pattern can be modeled with uncertainty, σ, as a decreasing function of the fixed cost. Specifically, we can assume

$$\sigma = \sigma_0 + e^{-lK}$$

where σ_0 is the base level of uncertainty, K is the fixed cost and $l > 0$ is a coefficient. Assume the unit product value is 1, discount rate is 5 % per annum, duration of project is 10 years. Assume σ_0 is 20 % per annum and l is 0.2. Calculated rate of return from the project with different levels of fixed cost is shown in Fig. 3.4. When the level of fixed cost is increased, the rate of return increases initially and then declines.

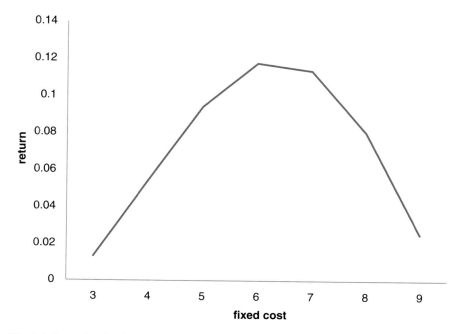

Fig. 3.4 Increasing fixed cost to reduce uncertainty, showing the relation between level of fixed cost investment and rate of return

Many decision makings involve the spending of fixed cost to reduce uncertainty, such as unemployment insurance, old age insurance, medical insurance, and governments' guarantee to financial institutions. People often take polar positions in debating these issues. A good quantitative model will help us reach compromise among various parties.

3.3.3   Resource Quality and Investment Patterns

One way to represent the resource quality is by the level of uncertainty rate. In physics, the term representing uncertainty rate in lognormal process is often called the diffusion rate, or the dissipation rate. A higher dissipation rate means that more energy is wasted as heat and less energy is applied to do useful work, indicating low quality of energy fuels. For example, when a dry cell gets discharged, its internal resistance gradually increases and more energy turns into unusable heat. The quality of the dry cell declines over time. So the quality of resources can be represented by the (inverse of the) uncertainty rate, or dissipation rate.

Table 3.1 lists the fixed costs, durations, net present values at different levels of dissipation rate when net present value from formula (3.13) is maximized with respect to fixed cost and duration, assuming a constant discount rate (4 %). Low dissipation rates, which represent the availability of high quality resources, correspond to the choice of high fixed costs and long durations, and to high net present values. Since fixed costs and durations are more visible than resource quality, highly valuable investments are often seen to be the result of those investment choices. More generally, high fixed costs and long durations are often associated with progress.

From Table 3.1, when resource quality is high (dissipation rate equals 0.2), the investment that generates the highest net present value has a fixed cost of 10.3 and duration of 42.8 years. The investment performance of such a level of fixed cost with different levels of resource quality can be calculated from (3.13). The results are presented in Table 3.2. When resource quality becomes progressively lower, net present values of the same investment decline and eventually drop below zero.

In the past several centuries, with continuous improvement in the extraction and use of fossil fuels, the amount of high quality resources consumed has increased steadily. Social institutions make many adjustments, generally increasing the fixed cost and duration of investments to take advantage of this abundance. Because of

Table 3.1 Investment decisions and values of investments with different levels of resource quality

Dissipation rate	0.2	0.25	0.3	0.35	0.4	0.45
Fixed cost	10.3	8.1	6.4	5.1	4.1	3.3
Duration	42.8	35.7	29.7	24.9	20.9	17.8
Net present value	18.4	13.6	10.1	7.6	5.7	4.3

Table 3.2 Values of high fixed cost, long duration investment with different levels of resource quality

Dissipation rate	0.2	0.25	0.3	0.35	0.4	0.45
Fixed cost	10.3	10.3	10.3	10.3	10.3	10.3
Duration	42.8	42.8	42.8	42.8	42.8	42.8
Net present value	18.4	13.0	8.3	4.1	0.6	−2.2

the high correlation between fixed cost, life span and general prosperity, decision makers often reflexively adopt policies that increase fixed costs and life span, expecting to take advantage of the greater efficiency those systems yield. Secondary education becomes mandatory and tertiary education becomes easily available in most wealthy countries, greatly increasing the fixed cost of social life. Retirement age has been continuously extended. However, in the last several decades, some social problems, such as high youth unemployment, below replacement fertility and population aging, have become increasingly prominent. Furthermore, the quality of available resources declines gradually. This is reflected from the steady decline of average EROI (energy return on investment) over the years. From our calculation, in an environment with declining resource quality, high fixed cost investment and long duration working years will lead to negative returns of social systems. From the calculation, it requires lower fixed cost and shorter duration to restore positive return. This means the reduction of fixed cost of the society and earlier, not later, retirement age. While it is politically difficult to implement these policies, the theory may help stimulate discussion on these important issues.

3.4   Monetary Policies and Business Cycles

Fischer Black once stated,

> I like the beauty and symmetry in Mr. Treynor's equilibrium models so much that I started designing them myself. I worked on models in several areas:
>
> Monetary theory
> Business cycles
> Options and warrants
>
> For 20 years, I have been struggling to show people the beauty in these models to pass on knowledge I received from Mr. Treynor.
>
> In monetary theory — the theory of how money is related to economic activity — I am still struggling. In business cycle theory — the theory of fluctuation in the economy — I am still struggling. In options and warrants, though, people see the beauty. (Mehrling 2005, p. 93)

Apparently, Black thought monetary theories, business cycles and options can be understood by related theories. His intuition was very insightful. In the following, we will apply the analytical theory just derived to understand monetary policies and business cycles.

The main tool of monetary policy is the setting of interest rate levels. We will examine investment decisions at countries with different levels of interest rate while keeping other parameters constant. Suppose in two countries A and B, annual output is 1 billion dollars. Uncertainty rate is 30 % per annum. Decision makers attempt to maximize the net present value of investment project, which is a common criterion in investment. Suppose discount rates are 3 and 10 % per annum in country A and B respectively. We will ask the following questions. How much will be the desired fixed costs and how long will be the expected project durations? What are the net present values of projects in countries A and B?

We maximize (3.13), which is net present value of an investment by changing fixed cost, K and duration, T when interest rates are set at 3 and 10 % per annum respectively. With commonly used software, such as Excel, we can find the solution easily. The following table lists the calculated results.

Interest rate (%)	Fixed cost	Duration	NPV
3	7.1	35.7	12.3
10	3.9	13.6	4.0

From the table, we find that when interest rate is lower, the amount of fixed investment is larger, investment duration is longer and the net present value is higher. So investors normally prefer low interest rate environment. However, the net present values are expected returns calculated at the beginning of a project. The actual returns depend on future environment. Suppose after the projects are built, the actual level of uncertainty becomes 80 % per annum instead of 30 % per annum as previously expected. We can recalculate the net present values from (3.13) to find the net present value of the first project, built in the low interest rate environment of 3 %, becomes −6.2 billion dollars while the net present value of the second project, built in the high interest rate environment of 10 %, becomes −2.0 billion dollars. Both projects suffer losses. But the first project suffers much more losses. When environmental conditions change, values of investment in low interest rate environment experience larger fluctuations. In other words, the monetary policy of low interest rate will generate greater business cycles. So business cycles are greatly tied to monetary policies. This theory provides a simple and clear understanding about the level of interest rate and magnitude of business cycles.

If low interest rate policy is associated with large magnitude of business cycles, why low interest rate policy has become the common choice for most governments most of the time? When interest rates are lower, the borrowing costs are lower. This is especially helpful for high fixed cost projects, which often require substantial financing. Higher fixed cost systems have lower variable cost and hence have more significant scale economy than lower fixed cost systems. When the market size for a product or a service is large and growing, a small number of large projects provide higher returns than a large number of small projects.

In a growing economy, businesses often anticipate growth. They build up very high fixed cost projects, and with their corresponding low variable costs, take advantage of future large market size. However, high fixed cost systems require higher levels of output to breakeven. While future demand is expected to be high, the current revenue is often not high enough to support the operation of high fixed cost systems. When there is a large gap between revenue and cost in an economic system, and the capital market is unable to fill the gap, recession occurs. To stimulate demand, governments often lower interest rate to ease borrowing from consumers. Hence, low interest rate policy has been used to stimulate both investment and consumption. Since high fixed cost systems are more competitive in a large and growing economy, and a low interest rate environment helps high fixed cost systems, low interest rate policy has in the past been adopted by most governments, despite their magnifying impacts on business cycles.

High fixed cost production systems, with their low variable cost, often reduce the average cost of products in a large and expanding economy. At the same time, the expansion of the economy tends to strain the availability of inputs, which often put an upward pressure on the costs of products. In the past, the abundance of cheap oil ensured that the inflation pressure was mild most of the time. However, with the increasing scarcity of resources, rapid economic expansions often trigger rapid commodity price inflations. After the burst of the internet bubble in 2000, interest rates were lowered to stimulate the economy. For several years, the housing market, stock market and commodity market rose continuously, which seemed to suggest that the low interest rate policy worked well in stimulating the economy. However, the rise of commodity prices became unstoppable. Eventually, high commodity prices and the subprime mortgage crisis, both a result of the low interest rate policy, generated the biggest recession since the Great Depression.

Balancing the need for economic growth and a check on inflation has always been a problem in implementing interest rate policies. In the past, governments mostly focused on the goal of economic growth by keeping the interest rate low. Low interest rate environments also generate high rates of inflation in stock and housing prices. But inflation in stock and housing prices is often given a positive interpretation. Inflation in the price of commodities, the raw material of all economic activity, had been kept low for a long time by a steady increase in commodity output. However, with the increasing physical cost of extracting many commodities, it becomes increasingly difficult to keep inflation in commodity prices low while simultaneously demanding a high rate of economic growth.

Low interest rate environments reduce financing cost to build up specialized production systems. It also reduces financing cost in purchasing generic scarce commodities. The scarce commodities could be raw materials, real estate in prime locations, or shares in companies with monopoly powers. Investment in production systems needs highly specialized skills and is very time consuming. The trading of production systems is highly illiquid. The markets for trading of generic commodities are highly liquid. When resources are abundant, there is little inflation pressure on the resource. To make a profit, companies have to make long-term investments in manufacturing. That was what happened to the world economy over most of the last several centuries. However, with the population growth and depletion of resources, many resources, such as land in prime locations, mineral reserves and agricultural products, have been under increasing inflation pressure. High inflation pressure in commodities increases investment opportunities in commodities and reduces investment opportunities in manufacturing, which relies on commodities as inputs. As a result, the investment capital has been migrating from specialized and illiquid manufacturing into generic and liquid commodities, real estate and financial assets. Overall, in an age of scarcity of commodities, low interest rate policy will be less and less able to stimulate economic activities and more and more able to stimulate speculative activities, generating waves after waves of economic instabilities.

The discount rate is a reflection of risk. In economic downturns, lenders naturally raise interest rates to compensate for higher risk. However, in economic downturns, central banks often lower interest rate to stimulate economy. Do not central banks and governments, or the general public, have to bear the same risks? They do! In response to the bursting of the stock market bubble in early 2000, the central banks lowered interest rates to stimulate the economy. The low interest rate environment inevitably stimulated speculation, especially highly leveraged speculation, which had a higher upside potential. The massive bailout of the major banks and other financial institutions indicated that governments and the general public, like other institutions or individuals, have to bear the same risks for low interest rate loans or guarantees. In the past, long term risk inherent in low interest policy was small for many countries because of the expected long term economic growth fueled by cheap resources. However, with increasing scarcity of resources, the prospect of long term economic growth in many countries is dim. Risk associated with the low interest rate policy will increase while the stimulus impact will decrease.

Under a low interest rate environment, financial institutions, which incur low borrowing costs, will benefit. Depositors, who receive low interest incomes, are harmed. The depositors represent the general public and the financial institutions represent a small minority of people. Why do low interest rate policies prevail for such a long time? This is because in a large and expanding economy, high fixed cost systems provide a higher return than lower fixed cost systems. A low interest rate environment makes high fixed cost systems more competitive. Successful companies create large number of high paying jobs. Part of the large amount of profits brought in by the high fixed cost large firms becomes tax revenues and is redistributed to the general public. Overall, in a large and expanding economy, a low interest rate policy benefits the general public despite their loss as depositors.

With a steady increase in the cost of extracting natural resources, however, steady economic growth has become increasingly difficult. In such economic conditions, the divergence of interest between financial institutions and the general public becomes a very important issue for social stability. In a no growth and declining economy, the problem of income distribution will become a very important political issue.

With a simple analytical theory, we show that interest rate levels impact not only the amount of economic output, but also the structure of economic system. As a result, interest rate levels impact not only short term economic activities, but also long term economic activities. With a simple analytical theory, we show very clearly about the relation between interest rate and spending. While each person's spending pattern differs, on average, there is a strong relation between interest rate and spending. Instead of demanding ordinary people stop borrowing, the central banks should stop the low interest rate policy. In a world of increasing cost of extracting natural resources, the continuation of low interest rate policy will generate wide gyration of social systems that we have witnessed in recent years. We are not proposing that central banks actively engage in a policy to maneuver interest rates. Instead, if the governments and central banks gradually reduce the magnitude of guaranteeing banks' assets and bailing out financial institutions, interest rate will rise to reflect the risk involved in financial transactions. Individuals and organizations borrow less when interest rate is high. This will reduce the occurrence and magnitude of financial crisis.

Hyman Minsky one said, "Stability is destabilizing". What does it mean exactly? With an analytical theory, we can obtain a very clear understanding from an example similar to the one at the beginning of this section. Suppose in two countries A and B, annual output is 1 billion dollars. Suppose interest rate is 5 % per annum in both countries. Uncertainty rate is 30 % per annum in country A and 55 % per annum in country B. Decision makers attempt to maximize the net present value of investment project. How much will be the desired fixed costs and how long will be the expected project durations? What are the net present values of projects in countries A and B?

We attempt to maximize (3.13), which is net present value of an investment by changing fixed cost, K and duration, T when uncertainty rates are set at 30 and 55 % per annum respectively. The following table lists the calculated results.

Uncertainty rate (%)	Fixed cost	Duration	NPV
30	5.8	25.3	8.5
55	2.1	12.1	2.3

From the table, we find that when uncertainty rate is lower, the amount of fixed investment is larger, investment duration is longer and the net present value is higher. So investors normally prefer low uncertainty rate environment. However, the net present values are only expected returns calculated at the beginning of a project. The actual returns depend on future environment. Suppose after the projects are built, the actual level of uncertainty becomes 80 % per annum in both countries due to reasons unforeseen by decision makers, such as a global financial crisis. We recalculate the net present values from (3.13) to find the net present value of the first project, built in the low uncertainty rate environment of 30 %, is −4.4 billion dollars while the net present value of the second project, built in the high uncertainty rate environment of 55 %, is 0.0 billion dollars. The first project suffers heavy losses while the second project barely breaks even. When environmental conditions change dramatically, values of investment in formerly stable environment experience large fluctuations. In other words, "Stability is destabilizing".

3.5 Equilibrium and Non-equilibrium Theory

Theories and terminologies are invented by humans to describe our world. In turn, they greatly influence our thinking.

General equilibrium theory forms the foundation of the established economic theories. From general equilibrium theory, wealthy countries have reached the desired optimal equilibrium state. So wealthy countries are called developed countries in economic literature, as well as in popular writings. Poor countries are certainly not in the desired optimal equilibrium state. But from the equilibrium theory, all economic systems, if allowed to operate in free market conditions, will always reach optimal equilibrium state. Hence poor countries are called developing countries.

Since wealthy countries are called developed countries, it is hard for most mainstream economists to imagine, let alone predict, that wealthy countries will fall into deep recessions. When deep recessions do occur in wealthy countries, given time, economies will inevitably return to the optimal growth path of the equilibrium state. Since poor countries are called developing countries, their conditions are expected to improve over time. When improvements don't occur, it must due to institutional deficiencies that distort market functions. Overall, global conditions will improve indefinitely, with only occasional temporary setbacks. Indeed, our time has been called the age of "Great Moderation". This is why our societies were unprepared for the 2007, 2008 financial crisis and are unprepared for future crises, judging from the policies we adopted.

From a non-equilibrium theory, wealthy countries, being further away from equilibrium state than poor countries, consume more resources to sustain themselves. This is consistent with empirical observation. When resources are abundant, wealthy countries, which have developed technologies and institutional structures to utilize large amount of resources, especially energy resources, are more powerful than poor countries. Wealthy countries, whose high fixed cost structures require stable environment, devote more resources to maintain stability. As a result, wealthy countries are more stable than poor countries most of the time. However, wealthy countries, due to their higher fixed cost, could suffer more than poor countries when they could not develop effective mechanisms to deal with specific instability. Poor countries have less technology to utilize large amount of energy and other resources to work for them. So they are less powerful than the wealthy countries most of the time. But simpler technologies also require fewer resources to maintain them. Poor countries can endure resource scarcity and unpredicted uncertainty better than wealthy countries.

Biologists have pondered similar questions.

The constant conditions which are maintained in the body might be termed equilibria. That word, however, has come to have fairly exact meaning as applied to relatively simple physic-chemical states, in closed systems, where known forces are balanced. The coordinated physiological processes which maintain most of the steady states in the organism are so complex and so peculiar to living beings — involving, as they may, the brain and nerves, the heart, lungs, kidney and spleen, all working cooperatively — that I have suggested a special designation for these states, homeostasis. The word does not imply something set and immobile, a stagnation. It means a condition — a condition which may vary, but which is relatively constant." (Cannon 1932, p. 25)

Cannon considered and rejected the term equilibrium in describing steady states in the internal environment of human bodies. He was aware his thoughts "would be suggestive for other kinds of organizations—even social and industrial". Indeed, biologists have recognized non-equilibrium theories provide better descriptions of living systems than equilibrium theories since the time of Darwin. "In the old system, each species was imagined to have been created according to some ideal type. Variation was just so much noise superimposed on the ideal type. After Darwin, the variation itself was seen as real and important, while the notion of an ideal type was recognized as a useless abstraction" (Trivers 1985, p. 22).

Today, the non-equilibrium theory of living systems has been widely accepted in biology and other scientific studies. In social sciences, non-equilibrium theories probably are read and discussed by many students and researchers privately. However, they have neither been accepted nor actively discussed publicly. Probably, non-equilibrium theories could not satisfy some emotional needs for human beings. A non-equilibrium theory tells us that life with material abundance is difficult to attain, and if attained, difficult to maintain for a long period of time. But there is a natural longing for good life. Various theories are developed to satisfy this longing. Generally, these theories promise good life in the heaven or in next life, such as in religions, or good life in the future, such as in communism. In essence, they promise a great product without having to deliver it. Furthermore, by separating heaven and earth, religions can provide realistic and practical advice on our current life. This is the key to long term success for many religions. General equilibrium theory, as an adaptive product from an age of affluence, promises not only better life in the distant future, but also great life right now and near future. This is the attraction of general equilibrium theory and standard theories in social sciences, despite their inconsistency with basic scientific principles.

A dynamic and non-equilibrium theory provides more precise description of our societies than the equilibrium theories. But equilibrium theories often precede non-equilibrium theories due to their conceptual simplicity. Furthermore, there is no chasm between equilibrium theory and non-equilibrium theory. Our production theory was inspired by the option pricing theory, which was developed as a dynamic equilibrium theory.

3.6   Physics, Mathematics and Predictability

There have been a lot of criticisms about establishing economic theory on the foundation of physical principles. But both Jevons and Walras, the main founders of neoclassical economics, were trained as natural scientists. "Jevons did so many things that it is difficult to classify him by occupation. ... from examination of his other works we are inclined to list him rather as a physicist who wrote extensively on economics." (Jaynes 2003, p. 316). Both Jevons and Walras attempted to establish economics on the foundation of physics. From *The Coal Question* and other works, it is clear that Jevons understood economics from the perspective of thermodynamics. However, the mathematical methods of non-equilibrium thermodynamics were not yet developed in Jevons and Walras' time. Jevons himself made it very clear that "I believe that dynamical branches of the Science of Economy may remain to be developed, on the consideration of which I have not at all entered." (Jevons 1871, p. vii). Our theory can be understood as mathematical formalization of Jevons and Walras' vision.

Furthermore our body, our mind and our environment are physical systems governed by the same physical laws. Cars move much faster and carry much heavier loads than humans because cars are designed to consume much more energy than humans. Several years ago, it was mandated in some countries that cars should burn energy grown from the fields. Food prices soared as humans have to compete with cars for available crops. Riots erupted in many regions because of the high food prices. When we understand humans and human societies as both physical and economical systems, we will be able to assess the consequences of our actions more accurately. There will be much less "unintended consequences".

This is a mathematical theory of human societies. It is often suggested that events in human societies are not conducted in controlled environment, such as laboratories. Mathematical theories, which have been successfully applied in natural sciences, cannot be applied with the same level of effectiveness in human societies. But planetary movements are not conducted in controlled environment. It was the attempt by Copernicus, Kepler, Newton and many other pioneers to describe planetary movements mathematically that modern science was born. Some people suggest that events are human societies are not repeatable. Mathematical theories, which are supposed to describe repeatable events, cannot be applied to prediction with high level of confidence. But mathematical theories of atomic bombs and hydrogen bombs were developed before even a single atomic bomb or hydrogen bomb was detonated. Indeed, the very designs of nuclear bombs were guided by the mathematical theories of nuclear bombs.

To me, good mathematical theories are low cost experimental science. An actual global financial crisis can cost trillions of dollars. But with a mathematical theory built on a solid foundation, it takes only several hours of calculation to understand the links between the types of policies and the magnitude of business cycles and financial crisis. However, the policies that generate economic instability over long term often stimulate economic activities over short term. A mathematical theory describing economic activities over time may make people more aware of the long term consequences of specific policies.

Some leading economists often claim that nobody could have predicted the coming of financial crisis every time a financial crisis strikes. In times of financial crisis and large monetary losses incurred by financial institutions, it is often said that social science is not an exact science like engineering. So we will look at an example of engineering. If you climb up a play set, you may find signs like

"Maximum capacity: 150 lbs". Does that mean if you are 151 lbs, the play set will collapse? From our own experience, we know the play set will not collapse. In general,

> When an engineer estimates the weights which a bridge or beam must support, or the pressures to which a boiler will be subjected, he does not provide merely for those stresses in building the structure. The engineer multiplies his estimates by three, six or even by twenty, in order to make the structure thoroughly reliable. The greater strength of the material, above that calculated as necessary, measures what is known as a "factor of safety."(Cannon 1932, p. 231)

By regulation, financial institutions are required to have their capital level above capital adequacy ratios. By law, many governments are required to restrict spending or deficit at certain percentages of income or GDP. In most cases, financial institutions and governments would keep their financial ratios barely above regulatory or legal requirements. Rarely any financial institutions or governments would strengthen their financial ratio "by three, six or even by twenty", in order to make their financial structure thoroughly reliable. It is the low level of "factor of safety" in financial institutions and governments, and not that social sciences are not exact that causes many financial crises and large monetary losses.

If the factor of safety in financial system is increased to the level of that in engineering system, the financial system will be as safe as engineering system. At the same time, the income levels of top earners in financial firms will be reduced to that of top earners in engineering firms. This is the most important reason why the factor of safety in financial system is low. In engineering system, the payoff for reducing the factor of safety is low. But the punishment for engineering accidents is high. People responsible for injuries or deaths from low quality engineering projects will face legal responsibilities. By contrast, the payoff from reducing the factor of safety in financial system is high. But there is virtually no punishment for causing financial crises. Despite the systematic frauds in financial institutions that caused the 2007, 2008 financial crisis, not a single person was legally responsible for the frauds. There has been a great shift of attitude towards financial frauds. Many people got jail terms for savings and loans frauds in 1980s, which were at much smaller scale than the frauds that led to the 2007, 2008 crisis (Black 2005).

There are many ways to increase the "factor of safety" or lower the leverage of our financial system. A natural method is to allow the interest rate to reflect the actual market risk. Currently, by implicitly and explicitly guarantee financial assets, governments lower the risks of financial transactions by transferring their risk to the government and hence to the general public. This produces the low interest environment, which encourages borrowing. Policymakers are aware that people will borrow heavily in a low interest rate environment. They often warn people against heavy borrowing. Yet they keep the interest rate low to stimulate the depressed economy. Suppose policymakers often warn the public against drug use. Yet they keep the drugs at very low cost to stimulate the demand from the depressed people. Do you think policymakers should bear any responsibility? Macroeconomics and microeconomics are generally taught in different courses. However, micro behaviors are influenced by macro policies. By lowering the interest rate to encourage heavy borrowing and at the same time warning public against heavy borrowing, policymakers hope to receive credit for short term economic recovery and assign blame to the general public for the long term economic turmoil. This is the heart of the problem in economic policy making.

By keeping social sciences "inexact", many people can take advantage and avoid blames of the murky situation. But some prominent economists think that economic theories are already too mathematical. Paul Krugman claimed that economists mistook beauty for truth. Then he managed to dig up two examples of the supposed beauty. One was CAPM theory, a beauty half century old. From his description of CAPM, he apparently forgot what CAPM was. Even Krugman, renowned for his mathematical prowess among economists, could not properly name two mathematical theories in economics. Indeed, he didn't find any beauty in mainstream economic literature. All he found were "impressive-looking mathematics" "gussied up with fancy equations", which are generously applied in research papers as cosmetics, to cover up their lack of substance. Statistics shows that the influence of theoretical works has dropped sharply in the last several decades. But the deterioration of mathematical standard is not limited to economic theory. From college admission to textbooks in many areas of natural science, the standard of mathematics has been declining over time. Mathematics is abandoned because it provides objectivity and clarity. Mathematics becomes a politically convenient target because there are few, instead of many, mathematical thinkers in social science.

3.7 Concluding Remarks

The mathematical theory of social and biological systems is derived from basic economic and physical principles. The basic result of the theory is an analytical formula of the variable cost as a function of product value, fixed cost, duration of investment, discount rate and uncertainty, which are the other major factors in economic systems. With this formula, we can calculate total expected costs and net value of a particular investment in specific environment. If the net value of the investment by a system is positive, the system will grow; otherwise, it will decline. This analytical theory provides a single observable quantity to measure the performance of various economic, social and biological systems. It also provide a systematic method to guide our decision making on major economic and social issues, such as setting the discount rate and determining the fixed cost of a project investment or the whole society. This is very different from the standard economic theory, which proposes to maximize utility, an unobservable quantity.

As an application, the theory is used to study monetary policies and business cycles. From calculation, when interest rate is low, investment with high fixed cost and long duration will generate high expected profit. At the same time, high fixed cost, long duration investments are more sensitive to the increase of uncertainties. So the low interest rate policy has the potential to stimulate economic growth and to increase amplitude of business cycles. In an environment of abundant resources, the stimulation impact is strong and the destabilization impact is weak. In an environment of scarce resources, the stimulation impact is weak and the destabilization impact is strong. Our analysis provides a clear understanding how policies and social environment influence not only short term economic activities but also structures of our society and long term economic activities. When we evaluate policies and social structures, we need to consider not only short term performances but also long term consequences.

In general, it provides simple quantitative simulations of many important relations in biological systems and in our society. It can be applied to very broad areas. Since the theory was developed, it has been applied to project investment, corporate finance, trade and migration, resource and social structures, language and cultures, evolutionary and institutional economics, fiscal and monetary policies, firm size and competitions, software development economics and many other problems.

The general theme of the theory is the tradeoff between the fixed cost and the variable cost. This tradeoff is ultimately based on the Carnot principle in thermodynamics: The higher the energy gradient becomes, the less energy is wasted as heat. But at the same time, it is more costly to build and maintain systems that can withstand higher energy gradient. While the Carnot principle specifies the upper limit of the energy efficiency, actual energy of a system can be lower. In the language of economics, a lower variable cost system requires higher fixed cost. However, a high fixed cost system does not necessarily deliver low variable cost. The analytical theory developed in this chapter shows that the expected variable cost depends not only on fixed cost, but also on duration of the project, discount rate, and the expected uncertainty of the environment during the lifetime of the project. There are more detailed relations among these major factors in specific systems. Some relations are explored in this chapter. Much more need to be done in the future.

Chapter 4
Languages and Cultures: An Economic Analysis

4.1 Introduction

Economic theories usually are applied to understand concrete social systems, such as businesses and nations. But they also can be applied to understand "soft" social systems, such as languages and cultures. Languages and cultures are media of transmitting information. Their value rests on their ability to lower variable costs in communication. Lower variable cost systems in general entail higher fixed costs. Each culture or language has evolved to adapt to the local environment. As environment changes, however, different cultural systems may fare differently under new conditions. We apply the economic theory developed in the last chapter to study the co-evolution of languages, cultures and social systems. Many problems about the evolution and diffusion of cultures and languages and how they co-evolve with the social and economic systems can be understood in a very consistent way.

There is a long debate about the impact of differences in languages on human societies (Whorf 1956; Pinker 1994; Devitt and Sterelny 1999). The ability to acquire languages is innate and universal (Chomsky 1988). Written languages, however, appeared very late in the history of human evolution and independently originated only in very few places (Diamond 1997). The discussion will be confined to how the differences in written languages are related to other aspects of human societies.

Words in English, an alphabetic language, are linear combinations of twenty-six letters. Characters in Chinese, a logographic language, are two-dimensional pictures. Several thousand of these complex characters need to be memorised before one can read articles reasonably well. Therefore Chinese is much more difficult to learn than English (Hanley et al. 1999). For the same reason, Chinese is more spatially compact and visually distinct than English. A Chinese document is much shorter than an English document of the same content. The speed of reading in Chinese is higher than in English (Lu and Zhang 1999). From an economic point of view, the fixed cost (learning a language) of Chinese, a logographic language, is

© Springer Science+Business Media New York 2016
J. Chen, *The Unity of Science and Economics*,
DOI 10.1007/978-1-4939-3466-9_4

high and the variable cost (using a language) of Chinese is low. The opposite is true for English, an alphabetic language. Its fixed cost is low and its variable cost is high.

The classification of cultures by context is analogous to the classification of languages by fixed cost. "In general, high context communication, in contrast to low context, is economical, fast, efficient and satisfying; however, time must be devoted to programming" (Hall 1977, p. 88). This means that high context cultures are of high fixed costs and low variable costs. The opposite is true for low context cultures. The lowest-context cultures are probably Northern European and North American cultures, while "China, Japan, and Korea are extremely high-context cultures" (Anderson 2000, p. 266, 267). It is easy to note the link between the context of a culture and the fixed cost of learning a written language. Chinese language is a logographic language. Japanese and Korean languages are mixture of alphabetic and logographic languages. The logographic composition of these languages makes them more difficult to learn than alphabetic languages. Since learning and using written languages are such important parts of our lives, and since written languages are relatively stable and written records accumulate over time, they have a strong influence on many aspects of cultures.

Cultures and languages are media of transmitting information, which is the reduction of entropy (Shannon 1948; Bennett 1988). The value of cultures (and languages) rests on their ability to lower variable costs in communication. Lower variable cost systems entail higher fixed costs. In a homogeneous and densely populated society, people share common background and engage in communications frequently. It pays to spend more time to build up the context of the culture to reduce the variable cost of communication. In a sparsely populated society or a society with members of diverse background, it is more economical to keep the fixed cost of a culture low. Each culture has evolved to adapt to the local environment. As environment changes, however, different cultural systems may fare differently under the new conditions.

The performance of an economic system with respect to its fixed cost and variable cost can be studied with the analytical framework presented in the last chapter. From this framework, one can derive that as fixed costs increase, variable costs decrease rapidly in a low uncertainty environment and decrease slowly in a high uncertainty environment. The main insight from this theory is the trade-off between efficiency of high fixed cost systems in a stable environment and flexibility of low fixed cost systems in a fast changing environment. One of the major purposes of the research on culture is to find out the relation between culture factors and economic development. However, the mixed statistical results often puzzle cultural researchers. One type of culture that is linked to high economic growth in one period often performs badly in another period (Hofstede 1980). With this analytic framework, it can be analyzed in a straightforward way. Lower context cultures have advantages in fast changing environments and higher context cultures have advantage in stable environments.

Cultures are often called multidimensional phenomena. Some of the primary dimensions of cultures are context, individuality, power distance and uncertainty avoidance (Hall 1977; Hofstede 1980). From our analytic framework, one can derive that all these dimensions are linked to one single factor, the fixed cost or context of a culture. Although cultures are often expressed in colorful ways, at core, they display highly consistent patterns. This is not surprising, for the function of culture is the same across different cultures, to reduce the cost of communication.

This analytical framework helps understand some of the long standing puzzles in linguistics and many problems in the evolution of language and societies. Why didn't the ancient Egyptian language take the "natural" step to evolve into an alphabetic language? (Diamond 1997, p. 226). Was the logographic Chinese writing created independently or diffused from somewhere else? (Diamond 1997, p. 231). Why are there so many low fixed cost alphabetic languages, but so few high fixed cost logographic languages, although all the earliest written languages were logographic? Is it a coincidence that Chinese, a logographic language, has the most native users? Why are most of the original Chinese characters, used three thousand years ago, still used today, with many of these characters having retained their original meaning, whilst over the same period of time, most alphabetic languages have changed considerably? Why was China the wealthiest country in the world during the long period of the stable agricultural society? Why could not China initiate industrial revolution despite its immense wealth? Why do democratic systems have a long tradition in the environment where alphabetic languages are used, while they rarely developed in regions that use logographic languages?

Many of these questions have been answered by many people in many different ways with various levels of confidence. However, this analytical theory developed in last chapter will answer all these and other questions in a very simple and consistent way. The simplicity of the answers is not accidental. This is because the language, cultural and economic activities, which are thermodynamic processes, are directly modeled with an analytical thermodynamic theory. For any language, cultural or economic system, if it can help its hosts to extract more low entropy resources from the environment than the amount that dissipates, it will expand. Otherwise, it will contract.

This chapter is organised as follows. In Sect. 4.2, we apply the economic theory to give a unifying analysis of different dimensions of cultures. In Sect. 4.3, we make a detailed analysis of the evolution of written languages and how they affect cultural and economic development. Section 4.4 concludes.

4.2   An Application to Cultural Analysis

The value of the fixed asset rests on its ability to reduce the variable cost in production or information transmission. In general, the variable costs decrease when the fixed costs are increased. However, the rate of decrease is a function of uncertainty. As the fixed cost increases, the variable costs decrease rapidly in a low uncertainty environment and decrease slowly in a high uncertainty environment (Fig. 3.1). So the payoff of building up the context of a culture is high in a stable environment. In a fast changing environment, low context culture, being more flexible and innovative, have more advantages.

Suppose the fixed cost of understanding a culture is K, the variable cost per unit volume of information is C from (3.10). The volume of information transmission is Q. Assume the value of each transmission is S. Then the total cost of information transmission is

$$K + QC$$

while the total value of the information transmitted is QS. The rate of return of the information transmission is

$$\frac{QS - (K + QC)}{K + QC} = \frac{QS}{K + QC} - 1$$

Figure 3.2 is the graphic representation of the above formula for different levels of the fixed costs or the contexts of cultures. Two properties can be observed from Fig. 3.2. First, it takes higher volume of communication for the high context culture to break even. So it is more economical to keep the fixed cost of a culture low in a sparsely populated society. Second, higher context cultures have lower variable costs. In a densely populated society, the volume of communication is high. The return of a high context culture is higher than a low context culture.

Cultures are often called multidimensional phenomena. Some of the primary dimensions of culture are context, individuality, power distance and uncertainty avoidance (Hall 1977; Hofstede 1980). We will discuss the relations between these dimensions with the above framework.

"In general, high context communication, in contrast to low context, is economical, fast, efficient and satisfying; however, time must be devoted to programming" (Hall 1977, p. 88). This means that high context cultures share large amount of common fixed assets among their members. Because of this, members in a high context culture will value collectivism more than those in a low context culture, who will value individualism more. From Fig. 3.2, the return curve is steeper in a high context culture than in a low context culture. Power and wealth is more unevenly distributed in a high context culture than in a low context culture, which means high context cultures have higher power distances. From Fig. 3.1, high fixed assets systems perform better in low uncertainty environment. So high context cultures entail a higher level of uncertainty avoidance than low context cultures.

Hofstede (1980) and Hall's (1977) works indicated that the values of these four dimensions of culture variation are positively correlated. Their correlation may even be higher than the statistical results because some formulas that calculated the indices may not represent the defined meanings adequately. For example, Singaporeans are generally considered very cautious. However, Singapore scored lowest on the Uncertainty Avoidance Index (Hofstede 1980, p. 165). This is because this index is derived partly from employment stability and Singapore has a high job turnover rate. In a small and densely populated city state such as Singapore, changing jobs cause very little uncertainty. First, changing jobs rarely requires changing homes. Second, there are many firms crowded in a small place. It is often easy to find similar jobs nearby.

This analytic framework also makes it easy to analyze the relation between cultural factors and economic growth. For example, Hofstede (1980, p. 205) found that the Uncertainty Avoidance Index is negatively correlated with economic growth in the volatile period from 1925 to 1950 while it is positively correlated with economic growth in the stable period from 1960 to 1970. From Fig. 3.1, a system high in fixed assets, (and hence a high need for uncertainty avoidance) performs well in a stable environment and performs poorly in a volatile environment.

Most cultural and economic indicators are highly correlated with latitude (Hall 1977; Hofstede 1980; Parker 2000). High latitude areas receive less solar energy. So bio-densities in general and human population densities in particular are lower in the cold high latitude areas than in low latitude areas, where there is abundant solar energy. In high density areas, people interact more. Therefore it is more efficient to develop high fixed cost communication systems that have lower variable costs. People in the warmer low latitude areas, such as Southern Europe, generally develop high context cultures. In low density areas, people interact less. It is more economical to keep the fixed cost associated with communication small. People in the cold high latitude areas, such as Northern Europe, generally develop low context cultures.

4.3   The Evolutionary Patterns of Languages

It is more intuitive to map what we see into pictures than into alphabets. All the independently created written systems, from Sumer, China and Mesoamerica, are logographic languages. "The first Sumerian writing signs were recognizable pictures of the object referred to. … The earliest Sumerian writing consisted of non-phonetic logograms." Gradually, phonetic representation was introduced to write an abstract noun "by means of a sign for a depictable noun that had the same phonetic pronunciation" (Diamond 1997, p. 220).

The first solution to a problem is often very complex. When a problem is known to be solvable, later solutions tend to be much simpler. As a writing system diffused to other regions, many logograms were gradually simplified to easy to write alphabets, especially the abstract nouns and grammatical items that are best represented phonetically instead of logographically. The earliest alphabets can be tracked to ancient Egyptian languages, which is probably influenced by the Sumerian language. But they kept the logograms in their language. This has puzzled some people. "The Egyptians never took the logical (to us) next step of discarding all their logograms …" (Diamond 1997, p. 226). However, if we look at the patterns of innovation, the puzzle can be easily resolved.

Logograms, which are difficult to learn, contain high information content. They are of high fixed cost and low variable cost. To the people who already learned the logographic languages and hence already invested on the fixed cost, there is little incentive to convert the logograms into alphabets. It is for the similar reason that Japanese language, which is influenced by logographic Chinese language, retain many of the logograms. Only when the writing systems diffused further into new environments, the low fixed cost alphabetic system were gradually established. Semites familiar with Egyptian languages discarded all logograms and transformed the language into a pure alphabetical language (Diamond 1997, pp. 226–227). This pattern is very similar to the innovation of new products. Many of the new ideas are initiated inside established large companies. However, it is often the new and small companies that implement the novel ideas.

The spread of writing systems is also an evolutionary process from logographic to alphabetic systems. This observation helps us resolves the puzzle whether the logographic Chinese writing was created independently or was diffused from Sumerian writing (Diamond 1997, p. 231). If the Chinese writing was diffused from Sumer, it would have been evolved into an alphabetical language over such a long distance and over such a long time.

Next we will apply this analytical framework to understand several properties of languages. First, since alphabetic languages are of low fixed costs, they are very easy to spread out. That is why there are many more people using alphabetic languages instead of logographic languages. Because of the simple structure of alphabetic languages, it is easy for them to mutate into new forms to adapt to local dialects. That is why there are many more alphabetic languages than logographic languages and most of the alphabetic languages have a relatively small number of users. Because of the high fixed cost, a logographic language is difficult to get established and to sustain itself. That is why there are few logographic languages left today although all the earliest written systems were logographic.

Why has the Chinese language survived while all other logographic languages were eventually replaced by alphabetic languages? "China's long east-west rivers … facilitated diffusion of crops and technology between the coast and inland, while its broad east-west expanse and relatively gentle terrain, … facilitate north-south exchanges. All these geographic factors contributed to the early cultural and political unification of China" (Diamond 1997, p. 331). The same geographic factors also help the Chinese language to spread out quickly and gain large number of users. A logographic language has to be used by many people so that the low variable cost can offset its formidable high fixed cost. Chinese, a logographic language, has more than one billion native users, which is the highest among all languages.

Second, alphabetic languages, with low fixed costs, can create new words easily and absorb words directly from other languages. Logographic languages, with high fixed costs, are more conservative and change less. Most of the characters Chinese used three thousand years ago are still used today, and many of these characters retain their original meanings. Over the same period of time, most alphabetic languages have changed considerably. New words in Chinese are formed by new combinations of existing characters. Since each Chinese character carries distinct meanings that are very stable, it is often difficult to create proper words to represent really novel ideas, which delays the understanding and adoption of new ideas. When the Japanese language absorbs a new word from other sources, the sound will first be represented by the alphabetic part of the language. If the word becomes popular, people will gradually create a logographic word with a distinct meaning to represent it for easier communication (Harigaya 2001).

Since societies using logographic languages are more stable, people there will value past experiences and are more conservative. Since societies using alphabetic languages change fast, a forward looking perspective is more valuable.

People using an alphabetic language spend less time learning the language, their common property. So they value individualism more. People using a logographic language spend far more time learning the language, their common property. So they value collectivism more. Hofstede (1980) found that those Chinese-majority regions "Taiwan, Hong Kong, and Singapore score considerably lower on individualism than the countries of the western world." He attributed it to a particular Chinese philosophy (pp. 215, 231). We find that language, by itself, offers a clearer explanation.

Third, economically, regions using logographic languages will do well in a stable environment because of the low variable cost in communication. During the stable agricultural society, which occupied most of the past two thousand years, China was the wealthiest country in the world. However, the high fixed cost Chinese language makes it difficult to initiate new changes. Instead, changes are initiated in the regions of diversified alphabetic languages. An alphabetic language, with low fixed costs, is more flexible and more innovative, which enjoys advantages in a fast changing environment. Regions using alphabetic languages have been leading the innovative changes since the beginning of industrial revolution several hundred years ago.

The difference in performance is not only reflected in world history but also in industries at different stages of maturity. Regions using alphabetic languages, such as USA, are dominant in new technologies. Regions using a mixture of alphabetic and logographic languages, such as Japan, absorb and perfect the advanced technologies. China, the region using logographic languages, is the main manufacturer in the mature industries.

Fourth, in a logographic language environment, if one manages to learn this language, one can acquire a huge amount of information at high speed. But the process of learning this language is difficult. Those who fail to master it may be left behind. The difference in payoff is drastic. In an alphabetic language environment, the result is not that drastic because of the low fixed cost, or low barriers of entry (Fig. 3.2). One would expect the logographic language environment to be more elitist and the alphabetic language environment to be more pluralistic . Indeed, democratic institutions have a long tradition in the environment where alphabetic languages are used, while they rarely developed in regions that use logographic languages.

4.4 Concluding Remarks

In this chapter, the analytical theory of economics is applied to languages and cultures to offer a unified understanding of the co-evolution of languages, cultures and social systems. Written languages and cultures, which are created by human beings, in turn, exert great influence on the path of the development of human societies.

Chapter 5
The Entropy Theory of Mind

5.1 Introduction

Before we discuss the details of human mind, we may reflect on a simple and obvious fact: the size of our brains is much smaller than the world we try to comprehend. This fact alone determines that we can only store and process a tiny fraction of information that is available in the world. We also work under a tight energy budget. The energy consumption of our brain is less than a typical light bulb that lights our rooms. For comparison, Google's search engines consume more electricity than a million typical households. The mind, as a product of biological evolution, is subject to the same economic principle that its cost must be lower than its value. It would not be economical for the mind to develop capacities to detect everything. Indeed, human beings have only limited capacities to detect many frequently occurring events. Our eyes can detect only very narrow ranges of electromagnetic waves. We don't have sense organs to detect electric fields, while some fish do. Our genes related to the sense of smell are highly degenerated. This suggests our ancestors could smell much better than we do. Dogs' smell is much more sensitive than human's. Proteins are constructed from twenty kinds of amino acids. But our taste buds can sense only one of them. Since it is costly to develop and maintain information processing capacity, only the most frequently occurring events that are highly relevant to our survival will be detected by our senses and processed by our mind.

To further develop the idea that the cost of our information processing has to be lower than the value of information, we must first find a proper measure of cost and value of information. In 1870s, James Maxwell, in a thought experiment, linked the cost and value of information processing to entropy. His argument is very simple. If the cost of obtaining information is less than the reduction of entropy of a physical system, then the second law of thermodynamics is violated. So the cost and value of

© Springer Science+Business Media New York 2016
J. Chen, *The Unity of Science and Economics*,
DOI 10.1007/978-1-4939-3466-9_5

information is closely linked to physical entropy. In 1948, Shannon defined information mathematically by the entropy function. As a result, major problems in information theory were resolved very easily. The entropy theory of information works so well in understanding real world problems in communication because entropy and related functions provide good measures of value and cost in information transmission under different conditions. In other words, the entropy theory of information works so well because it is a good economic theory of information.

The relation between information and entropy can be understood from another perspective. Among many functions of the mind, the most important one is to identify resources at low cost. All organisms need to obtain resources for survival. While the forms of resources are diverse, most resources can be understood from a unifying principle. A system has a tendency to move from a less probable state to a more probable state. This tendency of directional movement is what drives, among other things, living organisms. Intuitively, resources are something that are of low probability, or scarce. The measure of probability of a system is called entropy in physics. In a formal language, systems move from low entropy state to high entropy state. This is the second law of thermodynamics, the most universal law of nature. The second law is often understood from an equilibrium perspective, rendering entropy an image of waste and death. However, from the non-equilibrium perspective, the entropy flow, which is manifested as heat flow, light flow, electricity flow, water flow and many other forms, is the fountain of life. Since all living organisms need to tap into the entropy flow from the environment for survival, entropy is a natural measure for value of information.

Human beings, as social animals, need to communicate their power and attractiveness to influence the behavior of others, such as potential mates. The entropy law, which states that systems tend towards higher entropy spontaneously, is the most universal law of the nature. It is natural that the display of low entropy levels evolves as the universal signal of attractiveness and power among social animals. Large body size, colorful and highly complex feather patterns with large amount of information content and exotic structures, the creation of distinct art works, the demonstration of athletic prowess, the accumulation of wealth, and conspicuous consumption—all of which represent different forms of low entropy— are the major methods of advertising one's attractiveness and power.

The above discussion suggests a natural connection among entropy, information and human mind. Indeed, many pioneering works have been done to make such connections. But it is generally thought that objective measures, such as entropy, only have limited use in understanding mind because thinking is subjective. However, human mind has evolved to process information at low cost. Languages are a window to human mind. Patterns of languages often reveal how our minds process information. In languages, not all words are of the same length. In general, more frequently used words are shorter than less frequently used words. For instance, in the sentence, "I climb a mountain." the word "I" has only one letter and the word "mountain" has eight letters. This pattern develops because "I" is used much more frequently than "mountain". "Words get shortened as their usage becomes more common. Thus, taxi and cab came from taxicab, and cab in turn came from cabriolet" (Pierce 1980, p. 246). Automobile becomes car; bicycle becomes bike; television set becomes television and then simply TV; personal computer becomes PC. By representing high probability events with shorter expressions, we reduce the time and effort in information transmission. Therefore, language is not a purely random mapping from the concrete worlds to the abstract symbols. It is a highly structured coding system that reduces the average length of messages.

Since languages are highly structured coding systems, even after messages are enciphered, some hidden structures remain. It is through the detection of these hidden structures that people tried to decipher these messages. The modern information theory was largely born out of attempts to decipher encrypted messages (Beutelspacher 1994; Boone 2005). Many pioneers in information theory, such as Alan Turing, Claude Shannon and Solomon Kullback, were involved in war time effort in breaking enemy cryptosystems. It is interesting to note that many successful traders in the financial markets are originally trained in the field of information theory (Patterson 2010).

The above discussion about languages indicates that human thinking is subject to the same economic principle to achieve a task at low cost. The entropy theory of information works well in understanding human mind because it provides a good economic measure on the cost of thinking and psychological patterns, which are low cost methods of information processing.

There is a natural connection between human mind, emotion and thermodynamic laws.

Thermodynamics is … the science of 'desire'. The existence of atoms and molecules is dominated by 'attractions', 'repulsions', 'wants' and 'discharges', to the point that it becomes virtually impossible to write about chemistry without giving it to some sort of randy anthropomorphism. Molecules 'want' to lose or gain electrons; attract opposite charges; repulse similar charges; or cohabit with molecules with similar character. A chemical reaction happens spontaneously if all the molecular partners desire to participate; or they can be pressed to react unwillingly through greater force. And of course some molecules really want to react but find it hard to overcome their innate shyness. A little gentle flirtation might prompt a massive release of lust, a discharge of pure energy. (Lane 2010)

The rest of the chapter is organized as follows. Section 5.2 summarizes the main properties of the entropy theory of mind, in particular those that are most relevant to behavioral finance. Section 5.3 provides a reflection on the various aspects of the concept of entropy that are relevant to further discussion. Section 5.4 formally introduces the concept of entropy and its properties. Section 5.5 develops a general theory of innate psychological patterns and learning as means to reduce the cost of information processing. Section 5.6 presents the theory of judgment. Section 5.7 discusses how investors' judgments determine their trading decisions and the returns of their portfolios. Section 5.8 applies the entropy theory of mind to understand patterns of investor behaviors and security markets. Section 5.9 concludes.

5.2 Main Properties of the Entropy Theory of Mind

First, information is costly and the value of information is highly correlated with the cost of information. From James Maxwell (1871), if the physical cost of obtaining information is less than the reduction of entropy in a physical system, then the second law of thermodynamics is violated. Because he was confident the second law of thermodynamics is universal, the cost of obtaining information must be higher than the reduction of entropy in a physical system. In other words, the cost of information must be higher than the value of information. If this is true, why we even bother to obtain information? This is because some patterns in nature last for a long time and hence the same information can be used again and again. So the total value of certain information may be higher than the cost of obtaining the information. For example, the sun is hotter than the earth and probably will be for billions of years. As a result, the average frequency of light emitted from the sun is much higher than the average frequency of light emitted from the earth. In other words, the earth receives low entropy light from the sun and emits high entropy light toward space. The earliest organisms that successfully utilized this information with photosynthesis could use it again and again and replicate its genes to pass the information to their descendants. Information has positive value only when there is a persistent pattern related to that particular information. If some information has positive value, the carrier of that information will grow until other constraints reduce the net value of that information to zero. For example, since the earliest organisms developed the ability to absorb solar energy, their offspring spread all over the world to fill most places on the earth, until the constraints of available land and nutrients prevented their further expansion. At this time, the net value of photosynthesis of each plant approaches zero, or the cost of information on photosynthesis is close to the value of information on photosynthesis.

Different information has different value and cost. Only certain information has positive value to certain people at a certain time. We will pursue certain knowledge and neglect others. We also expend different amounts of effort to obtain different kinds of information. There are about five times more sensors of coldness than sensors of hotness under our skin. This is because humans have less capacity to adjust to coldness than to adjust to hotness. So detecting coldness is especially important to us for our survival. Although our fingers are much smaller than our back, the area in brain responsible for information processing from fingers is much larger than the area in brain responsible for information processing from our backs. There are twenty kinds of amino acids in nature. But our tongue can detect only one of them, glutamate. This is because glutamate is the most important amino acid in biochemical processes. Since different information has different values, we naturally assign different weights to different information. In other words, all of us are naturally biased. People often accuse others of being biased. However, being selective is the very essence of information processing.

In general, information of high economic value also exhibits f high economic costs. This result helps understand the systematic differences in the trading patterns of large and small investors. Depending on the value of assets under management, different investors will choose different methods of information gathering with different costs. Large investors are willing to pay a high cost to collect and analyze fundamental information. Small investors will spend less cost or effort on information gathering and rely mainly on easy to understand low cost information such as coverage from popular media and technical signals. Empirical works confirm that institutional investors trade on fundamental information while individual investors trade on price trends and news (Cohen et al. 2002; Barber and Odean 2008; Engelberg and Parsons 2011).

The differences in information processing by large and small investors generate the differences in their trading behaviors. There is a time lag between firm activities, such as R&D and project construction, and profit realization. By engaging in costly research, large investors are in a better position to estimate the values of new projects before they turn profitable and are better at separating long term components from short term fluctuations in earning data. Small investors, lacking detailed information on firm activities, have to rely on realized earning figures to assess firm values or observe the stock price movement to infer the trading activities of the informed. Since the stock transactions by individual investors are often triggered by public media, they sometimes are highly correlated (Barber et al. 2009b). On average, large investors buy at an earlier stage when stock prices are rising and sell at an earlier stage when stock prices are falling than the small investors (Hvidkjaer 2006). As a result, large investors as a group make money and small investors as a group lose money from their trading activities (Wermers 2000; Barber and Odean 2000; Cronqvist and Thaler 2004). Chen et al. (2000) documented that shares bought by mutual fund managers outperform shares they sold. Odean (1999) documented that the shares individual investors sold outperform the shares they bought. The heterogeneity of information processing and resulting trading activities by different investors is the main reason behind the observed patterns in the asset markets.

Second, the entropy theory of mind provides a simple mathematical model for a unified understanding of learning and human psychology. From the information theory, the cost of information processing depends on the relation between the structure of information sources and the structure of the coding system that transmit information. When the structure of coding system becomes more similar to the structure of the information sources that are to be transmitted, the average signal length will becomes lower. In other words, information processing is more efficient when the coding system represents the information sources more precisely. However, a more refined and specialized coding system performs poorly compared with a generic coding system when transmitting information without specific structures or with structures very different from the coding system. This tradeoff holds the key to understand human psychology and learning.

If certain events are common in the environment, it is economical to learn about them and represent them with shorter signals so mind can respond to them faster. When certain patterns persist for many generations, learning about these patterns is often transformed into more permanent structures in mind through epigenetic and genetic means so each generation does not have to relearn from scratch (Jablonka and Lamb 2006; Rando and Verstrepen 2007). These more permanent patterns of responses form the innate psychology. Learning and innate psychology complement each other. Learning is more costly but more flexible. Innate psychology is less costly but less flexible. Together, they provide us a coding system that lowers the average cost in information processing than an unstructured generic code in most situations that are important to us.

This integrated understanding of learning and human psychology will help us understand many patterns reported in behavioral finance literature and their evolution over time. Human psychology and past learning determine that decisions by investors in particular moments may not be optimal, especially with the benefit of hindsight. Learning also determines that a particular bias, if discovered and economically significant enough, will gradually reduce due to adaptation and competition. However, the learning processes can be complex and prolonged. For example, trend following has been a popular trading strategy for a long time. But the research on momentum has kept uncovering new and sometimes surprising quantitative results (Novy-Marx 2012). Furthermore, not all types of misevaluations of securities will decline overtime, since many misevaluations benefit major stakeholders who often are the best informed.

Third, the theory of judgment, a natural extension of the information theory, provides a quantitative measure of value and bias of our judgment. It also provides a link between our judgment and decision making. Kelly (1956) developed the link between information investors received and their trading decisions. In most time, people have to make subjective assessment of events without possessing complete information. The theory of judgment provides a measure to value one's judgment. The valuation of a judgment is against a reference state, which is usually taken to be the maximum entropy equilibrium state (Jaynes 1988). Since no additional information is required to determine the equilibrium state, the value of judgment from the decision making perspective can be naturally measured against the equilibrium state. However, the reference state can be a non-equilibrium steady state, such as a bubble state. Intuitively, if one buys a stock at two dollars and the equilibrium price is five dollars, then the value of your buying is three dollars. However, if the stock price can be momentarily moved to six dollars and you can take advantage of this high price, then the value of your buying is four dollars. Mathematically, the value of judgment is the average of profit or loss under different scenarios, which can be represented by a function generalized from relative entropy.

The value of judgment is always lower than or equal to the value of information with the same objective probability distribution and reference state. The value of judgment is equal to the value of information only when the subjective assessment of the probability distribution is identical to the objective probability distribution. Therefore, the concept of judgment is a generalization from the concept of information when a person does not have precise estimation of a random event, which is the case in most decision making processes. The difference between the values of judgment and information, or equivalently, the difference between actual cost of information processing and the lowest possible cost, is bias, which is defined by a mathematical function called relative entropy. Entropy and relative entropy are the two most important functions in information theory and statistical mechanics (Kullback 1959; Schlögl 1989; Qian 2001, 2009; Cover and Thomas 2006). Unlike the value of information, which is always positive, the value of judgment can be either positive or negative. This means that the value of a decision can be positive or negative. Applying to investment activities, this means that the value of active trading by investors can be either positive or negative. Trading that earn positive returns are generally attributed to information while trading that earn negative returns are generally attributed to behavioral biases. From the theory of judgment, the same judgment will have different values at different times due to changes of environmental parameters. Empirical evidences show that small individual investors often execute trades similar to those by large institutional investors but at a later stage. This could due to behavioral biases, or due to the difficulty of small investors to obtain timely information.

Under certain conditions, a judgment that is more biased may be more valuable than a less biased judgment. Suppose two investors are moderately favorable to two different stocks and each buys moderately amount of shares of respective stock. Subsequently, one stock performs very well and the other performs moderately well. Then the judgment of the investor who bought the stock that performs well is more biased. At the same time, the return from his investment is higher. This shows that value and bias of judgment are two distinct concepts. It will help clarify discussion in behavioral literature, which often identifies bias with low value of judgment. The theory of judgment bridges the chasm between the concept of information and cognitive bias. This will help provides a common framework for behavioral and informational perspectives in understanding financial market. By separating value and bias of judgment, we will also be able to measure judgments mainly on their values instead of biases. This will be especially helpful in measuring new ideas, which are often very biased but are also potentially very valuable.

Increasingly, our society measures people by their bias instead of value. A person accused of bias is often excluded from decision making processes. In political campaigns, a candidate accused of bias is often disowned by his political party and quickly dropped out of races. Being intolerant to bias often exclude the most valuable ideas as well. This is why the current public debate is often of little substance.

Traditionally, ideas are classified as rational and irrational. Our theory introduces the concepts of value and bias, which can be measured objectively. We hope these new concepts can allow us to discuss the merits of ideas more objectively.

Investment decisions are made according to investors' judgment about returns of different assets. To establish a precise link between investors' judgment and investment return, we consider a simple market with only two assets: a risk free asset and a risky asset. Based on the subjective assessment of the return distribution of the risky asset, an investor can determine the optimal portion of the risky asset in the portfolio and calculate the expected rate of return of this portfolio. We prove that the first order approximation of the expected rate of return of the portfolios constructed from a judgment is equal to the value of the same judgment. Therefore, the theory of judgment provides a quantitative link between the value of a judgment and the expected rate of return of the portfolio constructed from the same judgment. In a broader sense, the theory of judgment provides a link between ideas and their monetary values.

Since the judgment about the performance of a stock determines the level of holding about the stock, the change of judgment about a stock determines the volume of trading in the market, which is considered as the key ingredient missing from the asset pricing models (Banerjee and Kremer 2010). The theory of judgment provides a simple and intuitive tool to model trading volume in the asset market.

Many behavioral theories have been proposed to understand financial anomalies (Brav and Heaton 2002). Instead of developing a behavioral theory of economics directly, we propose an economic theory of behavior. Then we integrate the value and cost of information processing into the overall picture in economic decision making. The entropy theory of mind has been applied to understand many empirical patterns in behavioral finance (Chen 2003, 2004, 2007). The theory offers a simple and unified understanding of major patterns in market activities and investor behaviors.

5.3   The Concept of Entropy: A Reflection

More than any other scientific theory, thermodynamics was evolved directly from the attempts to improve the efficiency in human economic activities. Thermodynamics, and statistical mechanics, which is the micro foundation of thermodynamics, are the economic theory of nature and human societies. In the early days of research in thermodynamics, people focused on the quantities such as work and heat, which are intuitive to our senses. However, the amounts of work and heat transfer are path dependent, which makes it difficult to detect general patterns. In mid nineteenth century, Rudolf Clausius introduced the concept of entropy, which is a state variable and hence not path dependent. His work may remind us about path dependency in economics, which is often said to be the reason why economic activities cannot be described by a universal theory. In fact, researchers in science are faced with the same problems. The challenge is to find state variables, which are path independent, so the efficiencies of different path dependent activities can be compared.

Soon after the introduction of the concept of entropy, researchers recognized entropy as a fundamental quantity. Therefore it seems natural for entropy to be used as a coordinate in understanding thermodynamic problems. However, quantities such as volume and pressure, which are more intuitive to understand, are used as coordinates most of the time. Gibbs thought:

> The method in which the co-ordinates represent volume and pressure has a certain advantage in the simple and elementary character of the notions upon which it is based ... On the other hand, a method involving the notion of entropy, the very existence of which depends upon the second law of thermodynamics, will doubtless seem to many far-fetched,

and may repel beginners as obscure and difficult of comprehension. This inconvenience is perhaps more than counterbalanced by the advantages of a method which makes the second law of thermodynamics so prominent, and gives it so clear and elementary an expression. (Gibbs 1873a (1906), p. 11)

By using entropy as a coordinate, Gibbs was able to conceive the concept of free energy, one of the most important concepts in thermodynamics (Gibbs 1873b (1906)). However, current textbooks on thermodynamics continue to use volume and pressure as coordinate systems instead of entropy. This is a common pattern in many research works. The concept of entropy is crucial in developing many important scientific ideas, such as Planck's quantum theory and Einstein's photon theory (Planck 1900, 1901; Jammer 1966; Kragh 2000; Einstein 1905 (1989)). But in standard textbooks, the concept of entropy is rarely mentioned when introducing these ideas. This is because the concept of entropy, which cannot be directly measured by our sensory systems, is more difficult to communicate than the concepts of temperature, pressure, volume and speed, which can be directly accessed by our sensory system.

The above discussion shows the dual personality of entropy. It explains why the concept of entropy is behind so many scientific breakthroughs. It also explains why entropy is usually abandoned or its scope of application is restricted after the breakthroughs are established. The most prominent example is from Shannon himself. Shortly after Shannon (1948) developed the entropy theory of information, Weaver commented: "Thus when one meets the concept of entropy in communication theory, he has a right to be rather excited—a right to suspect that one has hold of something that may turn out to be basic and important" (Shannon and Weaver 1949, p. 13). This sense of excitement generated a lot of works to apply the concept of entropy to many other areas. As it is often the case, earlier attempts to apply some promising intuition do not yield concrete results easily. In an editorial, Shannon tried to discourage the jumping on the bandwagon:

Workers in other fields should realize that that the basic results of the subject are aimed at a very specific direction, a direction that is not necessarily relevant to such fields as psychology, economics, and other social sciences. Indeed, the hard core of information theory is essentially, a branch of mathematics, a strictly deductive system. (Shannon 1956)

It is true that the entropy theory of information is a strictly deductive system. But not every branch of deductive system achieves the same success in understanding information processing. So something else must account for its success. Entropy and related functions provide measures of value and cost in information transmission under different conditions. The entropy theory of information works so well because it is a good economic theory of information. Information technology has always been an extension of human mind in information processing. The principle of cost reduction in information processing applies to electronic devices as well as to brains, which are complex electronic devices built from neurons. Since entropy and related functions provide a good measure of information processing cost, it is natural that a theory that works so well on human-made information processing systems also work well on human information processing systems as well. From the economic perspective, there is really not much difference between the two. Indeed, Tooby, Cosmides and Barrett, three evolutionary psychologists, wrote a paper titled: "The second law of thermodynamics is the first law of psychology". We have shown in earlier works (Chen 2003, 2005) that many psychological patterns and many other problems in social sciences can be clearly understood from an entropy theory of human mind and decision making.

Since Shannon's statement, the entropy theory of information has been applied to many different fields. Many of these applications are now part of the standard textbooks in information theory, such as Cover and Thomas (2006). Nevertheless, due to Shannon's towering reputation, the academic fields have not been very receptive to the idea of an entropy theory of psychology, economics, and other social sciences. Today, Shannon's statement is still been invoked frequently to discredit attempts to apply the entropy theory of information to new fields. The best way to honor Shannon is to repeal his statement, which has not stood the test of time, so his original ideas can be applied to broader fields.

In the following, we will discuss the application of entropy theory of information in economic theories. Soon after Shannon developed the entropy theory of information, many attempts were made to give information more intuitive meaning. Kelly (1956) applied the information theory to investment decisions when investors have the knowledge of probability distribution of asset returns. He showed that the maximum exponential rate of return of an investor's asset is equal to the rate of information transmission. This result provides an economic meaning to the measure of information. Kelly suggested that the economic interpretation of information theory "can actually be used to analyze nearly any branch of human endeavor (Kelly 1956, p. 918)". Mathematically, the entropy based criterion is equivalent to maximization of geometric mean or logarithmic utility in portfolio investment (Latane 1959; McEnally 1986). The theory is very intuitive and practical. The entropy based investment theory has been successfully applied to investment practice (Thorp 1997; Poundstone 2005). Poundstone (2005) provided a detailed account on the interaction between the information theory based investment theory and academic economists. "Calling the Kelly system a "fallacy," he (Samuelson) helped persuade most economists to reject it (Poundstone 2005)". As Samuelson (1969) offered no substantial argument against the theory of Kelly and others, it is difficult for many to understand why he took such a strong view about it. But Samuelson's position was consistent with his general attitude toward the use of entropy in the social sciences:

> And I may add that the sign of a crank or half-baked speculator in the social sciences is his search for something in the social system that corresponds to the physicist's notion of "entropy". (Samuelson 1972, p. 450)

A distinct characteristic of the entropy based investment theory or geometric mean method is that it provides an objective criterion. In a scientific theory, an objective criterion is usually regarded as a good criterion. However, the conventional economic theories assume that human beings possess subjective utilities and subjective sensations, which make objective criterions less useful. So it will be helpful to investigate the concept of utility and sensation. Bernoulli's (1738 (1954)) paper is regarded as the first paper on utility. The utility function derived from his paper is a logarithmic function, which is equivalent to geometric return. The relation between subjective sensation and objective stimulation has been extensively investigated by Fechner. "If Fechner's psychophysics could be said to have an inverse square law, it was the logarithmic law relating sensation and stimulus (*Reiz*): $S = C \log R$" (Stigler 1986, p. 243).

In a section called *Can Perceptions Be Quantified?* Maor (1994) noted, "Among the many phenomena that follow a logarithmic scale, we should also mention the decibel scale of loudness, the brightness scale of stellar magnitude, and the Richter scale measuring intensity of earthquakes (p. 113)". The specific results about the relation between sensation and stimulus can be complex (Stevens 1961). However, empirical evidence overwhelmingly supports that subjective utilities and sensations are strongly related to objective stimulus and measurements. Furthermore, maximization of logarithmic utility or geometric mean is a dominant strategy under very general conditions (McEnally 1986; Sinn 2003). This is not to suggest that individual feelings and sensations are identical. The very opposite is true. We can discuss Newtonian mechanics as a parallel. All mechanical systems can be described by the basic laws of Newtonian mechanics. However, the movements of different mechanical systems can be very different and very complex. Because each person's neural system is unique by birth and individually calibrated by environmental stimulus, identical external stimulus may generate diverse sensations. What we suggest is that the entropy theory of mind provides a foundation to understand these complex sensations. In earlier works, we show that the entropy theory of mind provides a simple and consistent explanation to many psychological patterns (Chen 2003, 2005). Theories of mind from similar perspectives have been developed by other researchers (Jaynes 1988; Friston 2010). This shows the broad appeal of understanding the mind from physical and economic perspectives.

In economic research, the following often cited passage from Arrow has played a prominent role in defining the scope of application of entropy theory in the field of economic analysis.

The well-known Shannon measure which has been so useful in communications engineering is not in general appropriate for economic analysis because it gives no weight to the value of the information. If beforehand a large manufacturer regards it as equally likely whether the price of his product will go up or down, then learning which is true conveys no more information, in the Shannon sense, than observing the toss of a fair coin. (Arrow 1973)

The Shannon measure of information is

$$-\sum_i p_i \log p_i$$

It measures information value per symbol. If N is total number of symbols, the total information value is

$$N(-\sum_i p_i \log p_i)$$

So Shannon measure does give weight to the value of information. To further understand the weight issue, we will examine physical entropy directly. Shortly after the introduction of the concept of entropy, Boltzmann linked entropy, S, a macroscopic quantity, to W, the number of microscopic states with the formula

$$S = k \log W$$

We will derive the mathematical relation between the definitions of entropy by Boltzmann and Shannon, which will be important for us to understand the relation between information and physical entropy. Suppose a system has N particles, of which, n_i particles are in ith state, $i = 1, 2, \ldots, k$. Then

$$W = \frac{N!}{\prod n_i!}$$

where

$$N = \sum_i n_i$$

Taking the logarithm of W, we have

$$\log W = \log N! - \sum_i \log n_i!$$

Applying Stirling's theorem, we obtain

$$\log W = N \log N - N - (\sum_i n_i \log n_i - \sum_i n_i) = N \log N - \sum_i n_i \log n_i$$
$$= N(\log N - \sum_i \frac{n_i}{N} \log n_i)$$
$$= N(-\sum_i \frac{n_i}{N} \log \frac{n_i}{N})$$

Let

$$p_i = \frac{n_i}{N} \quad \text{for } i = 1, 2, \ldots k$$

Then

$$\log W = N\left(-\sum_i p_i \log p_i\right)$$

and

$$S = k \log W = kN\left(-\sum_i p_i \log p_i\right) \tag{5.1}$$

Since

$$-\sum_i p_i \log p_i$$

is defined as information by Shannon (1948), this establishes the mathematical relation between physical entropy and information. From (5.1), the physical and economic values of information are determined by three factors. The first factor is

$$-\sum_i p_i \log p_i$$

which is the Shannon information. It measures information value per symbol or per unit of output. The second factor is N. In physics, it represents the number of particles in a system. In information theory, it represents the number of symbols in information transmission. In economics, it represents the quantity of output. The third factor is k. It is a coefficient that converts value of one scale into another. In physics, it is the Boltzmann constant. In economics, it is the monetary value per unit of output. If the output is of high value, k is large. If the output is of low value, k is small. So entropy function does give weight to economic value of information.

It is not surprising that Arrow made such a simple mistake. Everyone makes simple mistakes. Many people must have spotted the error and some of them, such as Chen (2005), have pointed it out publicly. However, it is a telling sign that a statement containing such an obvious error has been cited or quoted as the final word on such important issue for so many years and anyone who points out the error is simply ignored.

5.4   An Introduction of Entropy Theory of Information

The value of information is a function of probability and must satisfy the following properties:

(a) The information value of two events is higher than the value of each of them.

(b) If two events are independent, the information value of the two events will be the sum of the two.
(c) The information value of any event is non-negative.

The only mathematical functions that satisfy all the above properties are of the form

$$H(P) = -\log_b P \qquad (5.2)$$

where H is the value of information, P is the probability associated with a given event and b is a positive constant (Applebaum 1996). Formula (5.2) represents the level of uncertainty. When a signal is received, there is a reduction of uncertainty, which is information.

Suppose a random event, X, has n discrete states, x_1, x_2, ..., x_n, each with probability p_1, p_2, ..., p_n. The information value of X is the average of information value of each state, that is

$$H(X) = -\sum_{j=1}^{n} p_j \log(p_j) \qquad (5.3)$$

The right hand side of (5.3) is the general formula for information (Shannon 1948).

The most important result from Shannon's entropy theory of information is the following formula

$$R(Y) = H(X) - H(X|Y) \qquad (5.4)$$

where $R(Y)$ is the amount of information one can receive, $H(X)$ is the amount of information a source sent and $H(X|Y)$, the conditional entropy, is called equivocation. Formula (5.4) shows that the amount of information one can receive would be equal to the amount of information sent minus the average rate of conditional entropy. Before Shannon's theory, it was impossible to assess accurately how much information one can receive from an information source. In communication theory, this formula is used to discuss how noises affect the efficiency of information transmission. But it can be understood from more general perspective. The level of conditional entropy $H(X|Y)$ is determined by the correlation between senders and receivers. When x and y are independent, $H(X|Y)$, $= H(X)$ and $R(Y) = 0$. No information can be transmitted between two objects that are independent of each other. When the correlation of x and y is equal to one, $H(X|Y) = 0$. No information loss occurs in transmission. In general, the amount of information one can receive from the source depends on the correlation between the two. The higher the correlation between the source and receiver, the more information can be transmitted.

The above discussion does not depend on the specific characteristics of senders and receivers of information. So it applies to human beings as well as technical communication equipment, which are the original focus in information theory in science and engineering. However, the laws that govern human activities, including mental activities, are the same physical laws that govern non-living systems.

$H(X|Y)$ in Formula (5.4) offers the quantitative measure of information asymmetry (Akerlof 1970). Since different people have different background knowledge about the same information, heterogeneity of opinion occurs naturally. To understand the value of a new product or new production system may take the investment public several years. To fully appreciate the scope of some technology change may take several decades. For example, the economic and social impacts of cars as personal transportation instruments and computers as personal communication instruments were only gradually realized over the path of several decades. This is one reason why individual stocks and whole stock markets often exhibit cycles of return of different lengths.

Because different people have different levels of understanding about certain information, this generate trading opportunities for the more informed.

> The global financial system may exist to bring borrowers and lenders together, but, over the past few decades, it has become something else, too: a tool for maximizing the number of encounters between the strong and weak, so that the one might exploit the other. Extremely smart traders inside Wall Street investment banks devise deeply unfair, diabolically complicated bets, and then send their sales forces out to scour the world for some idiot who will take the other side of those bets. (Lewis 2011, p. 153)

Actually effort to exploit information asymmetry is an ancient tradition. It has been widely used in trade, war and inside social organizations. In general, profit from information asymmetry is often called knowledge premium.

5.5   Learning and Psychological Patterns as Means of Reducing the Cost of Information Processing

Learning and psychological patterns are means of reducing the cost of information processing. For short term patterns, we often depend on learning to calibrate our mind temporarily. If certain patterns are stable in the environment for many generations, knowledge about them is often transformed into more permanent structures in mind through epigenetic and genetic means so each generation does not have to relearn from scratch at an additional cost (Jablonka and Lamb 2006; Rando and Verstrepen 2007). There is no dichotomy between innate psychology and learning, or between nature and nurture. Instead, they form a continuous spectrum in phenotypic plasticity. In general, the problems that occur often in life are easier to learn than problems that occur less often; the psychological patterns that evolve earlier from our ancestors are stronger than the psychological patterns that evolve more recently. Information theory and its generalization provide a quantitative framework to understand how coding systems affect the cost of information processing. We will begin with some general discussion and then present a numerical example.

Suppose a random variable, X, has n discrete states $\{x_1, x_2, \ldots, x_n\}$, with probability $\{p_1, \ldots p_n\}$. The lower bound for the cost of information transmission is the entropy of the random variable. Mathematically, the lower bound is

$$\sum_{j=1}^{n} p_j(-\log p_j)$$

where $\{p_1, \ldots p_n\}$ is the probability distribution of the random variable. The actual cost of information transmission by a coding system can be measured by

$$\sum_{j=1}^{n} p_j(-\log q_j) \tag{5.5}$$

where $\{p_1, \ldots p_n\}$ is the probability distribution of the random variable and $\{q_1, \ldots q_n\}$ is the subjective assessment of the probability distribution of the random variable when we design the coding system. Formula (5.5) may be called the generalized entropy function. It reaches minimum if and only if each

$$q_j = p_j, \ 1 \le j \le n$$

The purpose of encoding is to lower the cost of communication by accurately assessing the probability distribution of random variables. However, probability distributions of random variables may not be stationary. A coding system optimized from past data may not provide low cost communication in the future. To make the analysis more concrete, we will discuss a numerical example adapted from Shannon (1948). Suppose a random variable can generate four states, which are called state 1, 2, 3 and 4 respectively. A generic binary coding can be designed with the following mapping:

$$C(1) = 00$$
$$C(2) = 01$$
$$C(3) = 10$$
$$C(4) = 11$$

The coding has an average length per symbol of two. This corresponds to the maximum entropy of four random letters, which is

$$\frac{1}{4} \times (-\log_2 \frac{1}{4}) + \frac{1}{4} \times (-\log_2 \frac{1}{4}) + \frac{1}{4} \times (-\log_2 \frac{1}{4}) + \frac{1}{4} \times (-\log_2 \frac{1}{4}) = 2$$

Assume that, through experience, we learned that the probabilities of four states are

$$P(x = 1) = \frac{1}{2}$$

$$P(x = 2) = \frac{1}{4}$$

$$P(x = 3) = \frac{1}{8}$$

$$P(x = 4) = \frac{1}{8}$$

The entropy of the random variable with the above probability distribution is

$$H(x) = \frac{1}{2}\left(-\log_2\left(\frac{1}{2}\right)\right) + \frac{1}{4}\left(-\log_2\left(\frac{1}{4}\right)\right) + \frac{1}{8}\left(-\log_2\left(\frac{1}{8}\right)\right) + \frac{1}{8}\left(-\log_2\left(\frac{1}{8}\right)\right) = 1.75$$

From the information theory, the shortest possible average length of binary code per symbol can be as low as 1.75. Such a code exists with the following mapping.

$$C(1) = 0$$
$$C(2) = 10$$
$$C(3) = 110$$
$$C(4) = 111$$

We can confirm the average length of this coding by calculating

$$\frac{1}{2} \times 1 + \frac{1}{4} \times 2 + \frac{1}{8} \times 3 + \frac{1}{8} \times 3 = 1.75$$

This shows that learning can help us reduce the cost of information processing. From the information theory, in a code with shortest average length, the code length of events with probability p should be close to $-\log p$. In this case

$$-\log_2\left(\frac{1}{2}\right) = 1$$

$$-\log_2\left(\frac{1}{4}\right) = 2$$

$$-\log_2\left(\frac{1}{8}\right) = 3$$

The lengths of the codes designed are consistent with this criterion.

Now assume there is a fundamental change in the probability distribution of four states. The new probabilities become

$$P(x = 1) = \frac{1}{8}$$
$$P(x = 2) = \frac{1}{8}$$
$$P(x = 3) = \frac{1}{4}$$
$$P(x = 4) = \frac{1}{2}$$

If the coding system remains the same, the average length of code per symbol becomes

$$\frac{1}{8} \times 1 + \frac{1}{8} \times 2 + \frac{1}{4} \times 3 + \frac{1}{2} \times 3 = 2.625$$

which is longer than 2, the average length of generic code that can be designed without any knowledge about the probability distribution of the states. The average length of code per symbol also can be calculated from the generalized entropy function as

$$\frac{1}{8} \times (-\log_2 \frac{1}{2}) + \frac{1}{8} \times (-\log_2 \frac{1}{4}) + \frac{1}{4} \times (-\log_2 \frac{1}{8}) + \frac{1}{2} \times (-\log_2 \frac{1}{8}) = 2.625$$

which is the same as calculating average length of coding directly. From the above calculation, knowledge gained from past experience may hinder instead of helping information processing when environment change substantially or when we attempt new tasks that are very different from earlier experiences.

In this particular example, the cost reduction from a specialized coding system is $2 - 1.75 = 0.25$, when the specialized system provides an accurate representation of the world. When the world changes, the extra cost in information processing is $2.625 - 2 = 0.625$, which is much larger than 0.25, the previous cost reduction. This is a general pattern. It indicates that previous experience, expertise and knowledge can be very detrimental to adaptation in a changing environment. This explains why previous dominant species or societies are prone to failures while marginal species or societies come to dominance in new environment. In Chapter One and Three, it was shown that dominant, high fixed cost systems often have difficulty adapting to new environment while marginal, low fixed cost systems can adapt quickly. The approach there and the approach in this section generate similar conclusions.

The above discussion shows that the cost of information processing is low when a state with probability p is represented by a symbol of length about $-\log p$. When p is large, $-\log p$ is small. "In general, instances of large classes are recalled better and faster than instances of less frequent classes; that likely occurrences are easier to imagine than unlikely ones; and that the associative connections between events are strengthened when the events frequently co-occur" (Tversky and Kahneman 1974, p. 1128). A coding system that is a good representative of the information source has a lower cost of information processing than a generic coding system which does not require specific knowledge of the information sources. According to Shannon, "The transducer which does the encoding should match the source to the channel in a statistical sense (Shannon and Weaver 1949, p. 31)". Therefore it is economical to learn about the environment and develop coding systems accordingly. However, a more refined and specialized coding system performs poorly compared with a generic coding system when transmitting information without specific structures or with structures very different from the coding system. The choice of coding systems and their life spans depends on how persistent the environments are.

"In general, ideal or near ideal encoding requires a long delay in the transmitter and receiver (Shannon and Weaver, p. 31)". The exact length of delay for ideal encoding depends on the persistence of statistical patterns. The requirement for delay in encoding presents a tradeoff between efficiency and flexibility. For coding to be efficient, a long delay is required to accurately measure the statistical distribution of random processes. However, random processes are not always stationary. The longer delay in encoding, the slower a system responds to structural changes in random processes. In statistics, this corresponds to the tradeoff between type I and type II errors. An attempt to reduce one type of error will increase another type of error. In human psychology, this corresponds to conservatism and overreaction. An attempt to reduce one type of bias will increase the likelihood of another bias. In light of this understanding, should we continue to call these psychological patterns biases?

5.6 Value and Bias of Judgment

Most of the time, we have to make judgment and decisions without complete understanding of the information. Furthermore, the patterns we observe will change overtime. How can we measure the value and bias of our judgment, which we make based on the information we receive?

From the last section, three classes of probability measure are important to value decision making. The first is objective probability measure to particular types of events. The second is subjective probability measure by decision makers. The third is reference probability measure. We will construct measures of value and bias of our judgment from these three probability measures.

Suppose a random variable, X, has n discrete states $\{x_1, x_2, \ldots, x_n\}$, with probability $\{p_1, \ldots p_n\}$. The subjective judgment of a person may differ from the objective probability. Suppose the subjective judgment of the probability distribution is $\{q_1, \ldots q_n\}$, then the level of uncertainty of judgment on each q_i, is

$$H(q_i) = -\ln q_i \quad \text{for} \quad 1 \leq i \leq n$$

Here we use natural logarithm instead of logarithm based 2. It is easier to perform mathematical operations with natural logarithm. But they are equivalent. The total uncertainty of judgment of a random event is the average of uncertainty of judgment of each state, weighted by the objective probability distribution of the random event.

$$\sum_{j=1}^{n} p_j(-\ln q_j)$$

We will compare the uncertainty of the judgment against that of a probability distribution of a reference state. The reference state can be the equilibrium state, or a non-equilibrium steady state. Take a simple example of binary states of up and down in the stock market. Let $\{p, 1 - p\}$ represent the probability of up and down of market in the next period. If on average, stocks are up 55 % of the time and down 45 % of the time, $\{0.55, 0.45\}$ represent the equilibrium state and $\{0.3, 0.7\}$ represent a non-equilibrium state, which has a higher probability to go down rather than to go up in the next period. Suppose the reference probability distribution of a random event is $\{r_1, \dots r_n\}$. Then the total level of uncertainty of the reference state is

$$\sum_{j=1}^{n} p_j(-\ln r_j)$$

The value of judgment can be defined as the reduction of uncertainty from the reference state, which is

$$V(p, q, r) = \sum_{j=1}^{n} p_j(-\ln r_j) - \sum_{j=1}^{n} p_j(-\ln q_j) = \sum_{j=1}^{n} p_j(\ln \frac{q_j}{r_j}) \qquad (5.6)$$

The right hand side of Formula (5.6) is a function generalized from relative entropy. We will call it generalized relative entropy. When each

$$q_j = p_j, \qquad 1 \le j \le n$$

The value of judgment becomes the value of information.

$$\sum_{j=1}^{n} p_j(\ln \frac{p_j}{r_j}) \qquad (5.7)$$

From Gibbs inequality (Gibbs 1902),

$$\sum_{j=1}^{n} p_j \ln(p_j) \ge \sum_{j=1}^{n} p_j \ln(q_j)$$

Therefore, the value of judgment is always less than or equal to the value of information with the same probability distribution and reference distribution. The difference between the objective distribution and one's judgment, or the difference

between the values of information and judgment, is the measure of bias, which can
be defined as

$$B(p,q) = \sum_{j=1}^{n} p_j \ln(p_j) - \sum_{j=1}^{n} p_j \ln(q_j) = \sum_{j=1}^{n} p_j \ln\frac{p_j}{q_j} \tag{5.8}$$

which is the relative entropy function. It is always nonnegative and is zero if and
only if each

$$q_j = p_j, \ 1 \le j \le n$$

In general, the bias will be smaller when q_j is closer to p_j.

The value of judgment, the value of information and the measure of bias are
related by the following equation

$$\sum_{j=1}^{n} p_j (\ln\frac{q_j}{r_j}) = \sum_{j=1}^{n} p_j (\ln\frac{p_j}{r_j}) - \sum_{j=1}^{n} p_j \ln\frac{p_j}{q_j} \tag{5.9}$$

This equation shows that the value of judgment is equal to the value of information minus the measure of bias. The value of information is always positive. But
the value of judgment can be positive or negative. This means that active trading by
investors can increase or decrease the value of their investment portfolios.

In practice the reference probability distribution $\{r_1, \ ... \ r_n\}$ is often understood
as the maximum entropy distribution under known constraints. When there is no
known constraints, the maximum entropy distribution is $\{1/n, \ ... \ 1/n\}$ and Formula
(5.6) becomes

$$V(p,q) = \sum_{j=1}^{n} p_j \ln q_j + \ln n \tag{5.10}$$

We will perform some simple calculations to illustrate the properties of the value
of judgment and the measure of bias. For the simplicity of exposition, we will only
consider events with two possible outcomes, state 1 and state 2, in the remaining
part of the chapter. Intuitively, we can think state 1 as the up market and state 2 as
the down market. First, we will assume the reference state is the equilibrium state,
which is set to be $\{0.55, 0.45\}$. This means that it is more likely for the market to be
up than down. We begin with the calculation of the value of a judgment that is the
same as the equilibrium state. From (5.6), the value of the judgment that agrees with
the equilibrium state is

$$p \ln\frac{0.55}{0.55} + (1-p) \ln\frac{0.45}{0.45} = 0$$

Hence the value of the judgment that agrees with the equilibrium state is zero,
regardless of the actual probability distributions of the states. Intuitively speaking,

an investor who agrees with the market does not believe he or she possesses valuable information and puts his or her money into an index fund.

Now consider two random events with different probability distributions. Assume in the first event, the objective probability of state 1 is 65 % and the probability of state 2 is 35 %. Someone estimates the probability of state 1 is 57.5 % and the probability of state 2 is 42.5 %. We assume the equilibrium state remains {0.55, 0.45}. From (5.6), the value of his judgment is

$$0.65 \ln \frac{0.575}{0.55} + 0.35 \ln \frac{0.425}{0.45} = 0.0089$$

From (6), the bias of this judgment is

$$0.65 \ln \frac{0.65}{0.575} + 0.35 \ln \frac{0.35}{0.425} = 0.0117$$

In the second event, the probability of state 1 is 57.5 % and the probability of state 2 is 42.5 %. Someone estimates the probability of state 1 is 57.5 % and the probability of state 2 is 42.5 %. From (5.6), the value of his judgment is

$$0.575 \ln \frac{0.575}{0.55} + 0.425 \ln \frac{0.425}{0.45} = 0.0013$$

while the bias of this judgment is zero. From the above calculation, the judgment that is more biased can be more valuable than a less biased judgment under certain conditions. Intuitively speaking, an investor who is moderately favorable to a stock which turns out to earn very high rate of return will perform better than an investor who is moderately favorable to a stock which turns out to earn moderately rate of return. By separating value and bias of judgment, we will be able to perform more precise analysis to investor behaviors.

Next, we will consider values of judgments when reference states are equilibrium and non-equilibrium states respectively. Intuitively, we are comparing values of investment decisions when market settles down in equilibrium state or moves into a bubble state. Suppose an investor spot a good stock with high growth potential. Assume the objective probability of this stock to move up and down is {0.6, 0.4}. The investor's own assessment of the stock is {0.575, 0.425}. We will calculate the value of his judgment if the stock settles into the equilibrium state of {0.55, 0.45} or a bubble state {0.40, 0.60}, which means that the stock will have 40 % chance going up and 60 % chance going down the next time period.

When the stock will settles into the equilibrium state, the value of judgment is

$$0.6 \ln \left(\frac{0.575}{0.55} \right) + 0.4 \ln \left(\frac{0.425}{0.45} \right) = 0.0038$$

When the stock will move into the bubble state, the value of the same judgment is

$$0.6\ln(\frac{0.575}{0.4}) + 0.4\ln(\frac{0.425}{0.6}) = 0.0798$$

The value of the judgment in a bubble state is much higher than the value of the same judgment in the equilibrium state. Intuitively, investors holding shares of a stock benefit from the high stock price.

There is much discussion about why misvaluation occurs and why arbitrage cannot eliminate misvaluation. Most of the discussion is based on the assumption of psychological biases of investors, especially small investors. Empirical evidence shows that small investors are late-stage momentum traders, following the trends generated by large, informed investors (Hvidkjaer 2006). The systematic biases generated by small investors are of small magnitude and are reversed over a very short period of time (Hvidkjaer 2008; Barber et al. 2009a, b). However, many large scale profitable arbitrage activities, such as merger and acquisitions (Shleifer and Vishny 2003), require asset prices to be highly overvalued over extended periods. Most of the well informed about an asset are major stakeholders of the same asset. They often have an incentive as well as capacity to influence public opinion to move asset prices along certain direction. From the theory of judgment, the value of judgment and hence the value of investment depends on the reference state. Therefore the informed may choose to sell an overpriced asset to realize an arbitrage profit or push asset prices even higher to generate higher arbitrage profits. The above calculations indicate that it is much more valuable to generate asset bubbles than to realize profit immediately. Very often, it is easier for large stakeholders to profit from a big bubble instead of realizing a small arbitrage profit from a small bubble. For example, it was much more profitable and much easier for the major shareholders and top executives of AOL to merge with Time Warner at the peak of the internet bubble instead of selling its shares when AOL was only slightly overvalued.

The same logic applies not only in financial market but also in other social activities. Suppose a group of researchers work on a theory for a long time and get established for their works. If some of them recognize there are fundamental flaws in the theory, they could speak up against the theory and have their own reputation and their colleagues' reputation demolished. Or they could keep on advancing the theory and advancing their own careers and the careers of their friends. The choice is obvious. This is why bubbles are not a minor issue in financial market or social market. They have occurred and will occur repeatedly.

More systematic discussion on the properties of the theory of judgment can be found in Chen (2008). In the next section, we will discuss the link between investors' judgment, their trading decisions, and the expected rate of return of their portfolios. This will establish a relation between value of judgment and a monetary value.

5.7   Value of Judgment, Investors' Trading Decisions, and Expected Rate of Return of Portfolios

Investment decisions are made according to investors' judgment about stocks. To quantify the relation between investors' judgment and their trading decisions, we will consider a simple market with a risk free asset and a risky asset. When interest rate is measured on the inflation adjusted basis, the risk free interest rate can be set to be zero as a good approximation to reality. Here we will assume the return of the risk free asset to be zero. More general cases can be found at Chen (2011). The payoffs of one unit risky asset can be either $1 + d$ with probability p or $1 - d$ with probability $1 - p$. Investors can only assess the probabilities subjectively.

Investors aim at maximizing expected geometric return (Kelly 1956; Latane and Tuttle 1967). Based on the subjective assessment of the return distribution of the risky asset, an investor determines the optimal combination of the risk free asset and the risky asset in the portfolio. Then he or she can calculate the expected rate of return of this portfolio. Suppose an investor assesses the return distribution of the risky asset to be $\{q, 1 - q\}$. Assume the portfolio he or she constructed contains a portion x of risky asset and the remaining portion of $1 - x$ is risk free asset. The expected geometric return of the portfolio is

$$
\begin{aligned}
&((1 - x) + x(1 + d))^q ((1 - x) + x(1 - d))^{1-q} - 1 \\
&= (1 + xd)^q (1 - xd)^{1-q} - 1
\end{aligned}
\tag{5.11}
$$

To determine the value of x at which the portfolio will have the maximal rate of return, we differentiate the above formula with respect to x.

$$
\begin{aligned}
&\frac{d}{dx} ((1 + xd)^q (1 - xd)^{1-q} - 1) \\
&= d(1 + xd)^{q-1}(1 - xd)^{-q}(2q - 1 - xd)
\end{aligned}
$$

The above differentiation equals zero when

$$
x = \frac{2q - 1}{d}
\tag{5.12}
$$

At this value of x, the portfolio obtains the highest expected geometric return. Plug the value of x into (5.11), the expected rate of return is

$$
\begin{aligned}
&(1 + 2q - 1)^q (1 - (2q - 1))^{1-q} - 1 \\
&= 2q^q (1 - q)^{1-q} - 1
\end{aligned}
$$

If the objective return distribution of the risky asset is $\{p, 1 - p\}$ instead of the subjectively assessed $\{q, 1 - q\}$, the expected rate of return of the portfolio is

$$2q^p(1-q)^{1-p} - 1 \tag{5.13}$$

The first order approximation of (5.13) is

$$\ln(2q^p(1-q)^{1-p})$$
$$= p\ln q + (1-p)\ln(1-q) + \ln 2$$

Comparing the above result with (5.10), we find that the first order approximation of the expected rate of return of the portfolio constructed from a certain judgment is exactly equal to the value of the judgment. This result is an extension from Kelly (1956), in which investors are assumed to have precise information. As the value of judgment provides a good approximation to the rate of return on investment, it can be conveniently used to understand the relation between human judgment and patterns in investment returns and stock market.

From (5.12), the judgment about a stock determines the level of holding about the stock. The change of judgment about a stock determines the volume of trading in the market, which is considered as the key ingredient missing from the asset pricing models (Banerjee and Kremer 2010). The theory of judgment provides a link between investors' judgment and trading volume in the asset market, which will be applied to understand cycles of trading in the next section. In the following, we will apply the theory to calculate several numerical examples on the level of holding of risky assets.

The payoffs of one unit risky asset can be either $1 + d$ with probability p or $1 - d$ with probability $1 - p$. We can calibrate the equilibrium value of p and d with the empirical data on return and standard deviation. The arithmetic mean rate of return of the risky asset is

$$pd + (1-p)(-d) = (2p-1)d \tag{5.14}$$

and the standard deviation of the risky asset is

$$\{p[pd - (2p-1)d]^2 + (1-p)[(1-p)(-d) - (2p-1)d]^2\}^{1/2} = 2d\sqrt{p(1-p)} \tag{5.15}$$

respectively. The above relations show that there is a one to one correspondence between (p, d) and (mean, variance). The numerical representation by (p, d), just like mean and variance, is a simplified characteristic on the movements of the risky asset. While we assume a risky asset makes only two discrete movements with corresponding probabilities, in reality, a risky asset can make many different movements with corresponding probabilities.

Different people at different times in different places may have different opinions about the future of the stock markets. We will calculate the proportions of asset to be allocated to the risky asset with different expectations. When p is equal to 55,

57.5, 60, 62.5, 65 % while keeping d fixed at 0.25, the optimal allocations to risky asset, following formula (5.12), are

$$\frac{2p-1}{d} = \frac{2 \times 0.55 - 1}{0.25} = 0.4$$
$$\frac{2 \times 0.575 - 1}{0.25} = 0.6$$
$$\frac{2 \times 0.60 - 1}{0.25} = 0.8 \qquad\qquad (5.16)$$
$$\frac{2 \times 0.625 - 1}{0.25} = 1.0$$
$$\frac{2 \times 0.65 - 1}{0.25} = 1.2$$

while the arithmetic means of the risky asset, following formula (5.14), are

$$(2 \times 0.55 - 1) \times 0.25 = 2.5 \%$$
$$(2 \times 0.575 - 1) \times 0.25 = 3.75 \%$$
$$(2 \times 0.6 - 1) \times 0.25 = 5 \%$$
$$(2 \times 0.625 - 1) \times 0.25 = 6.25 \%$$
$$(2 \times 0.65 - 1) \times 0.25 = 7.5 \%$$

The real returns of the best performing stock markets in the world, such as United States, over the second half of the last century are close to or above 6.25 %. This justifies the standard statement of high risk, high return and the common practice of allocating most or all assets in risky securities in long term investments. When the arithmetic rate of return is above 6.25 % per annum, even higher rates of return can be achieved by borrowing funds to invest in risky assets, as calculated from (5.16). However, if the future expected real returns of risky assets are lower, the proportions of risky assets in investment portfolios should be lower as well to achieve maximum expected rate of returns (McEnally 1986).

We will use an example to illustrate the relation between risk and return of portfolios with different weights of risky asset. Suppose p is equal to 0.6 and d is equal to 0.25 for the risky asset. In three portfolios with risky assets at 60, 80 and 100 %, the geometric returns, calculated from (5.11), are

Proportion of risky asset	0.6	0.8	1
Geometric return of the portfolio	0.01903	0.02034	0.01899

The volatility of return of the portfolios will increase with the proportion of the risky asset. But when the weights of risky assets are higher than 80 %, the geometric return will decline, as shown from the above table. However, this does not mean all investors will make identical investment decisions. Different people have different assessment about the future market movements. Hence different investors may hold different portfolios.

In our discussion about decision making in investment, we use geometric rate of return. Return can be measured as geometric return or arithmetic return. Intuitively, return is better measured as geometric return, or compound return. However, in economic theory, arithmetic return is often used. For example, in CAPM theory, return is measured as arithmetic return. When return is measured as arithmetic return, decision makers need to balance the tradeoff between return and risk. Utility functions are introduced into economic theories to determine the exact tradeoff. Since the utility function is difficult if not impossible to measure for each individual, a theory based on utility essentially gives up any hope for describing reality accurately. When return is measured as geometric return, we simply need to maximize this return. There is no need to introduce utility function into decision making process. A theory based on geometric rate of return provides great details about decision making by investors. We can compare our theoretical predictions with reality. As a result, we can continue to improve our theory when our predictions are different from empirical observations.

5.8   An Application to Behavioral Finance

In this section, we will apply the theory of judgment to build a simple model to study trading behaviors of the heterogeneous investors and the resulting market patterns. It is an application to behavioral finance. Readers who are not engaged in the research of behavioral finance may wish to skip this section.

First we will determine the statistical distribution of investors with different levels of wealth. From earlier studies, such as those in Chatterjee et al. (2005), wealth distribution follows exponential law as a first approximation. Investors can be classified based on their wealth. Suppose each investor in group i has i unit of wealth. Since the number of investors in each group of wealth follows the exponential law, the proportion of investors with i unit of wealth is

$$\frac{1}{2^i}$$

Since

$$\sum_{i=1}^{\infty} \frac{1}{2^i} = 1$$

the proportion of investor population is normalized. The total wealth of the economy is

$$\sum_{i=1}^{\infty} \frac{i}{2^i} = 2$$

Because the investor population is normalized, the average wealth of an investor is 2.

From calculation, the Gini coefficient of this model economy is 33.3. The Gini coefficients of industrialized economies are roughly between 25 and 45. So the wealth distribution in this model economy is representative of real economies.

To further simplify discussion, we lump investors into three groups. The representative wealth for each member of the three groups of investors are 1, 4 and 9 respectively. The proportions of three groups of investors are determined by maximum entropy principle (Jaynes 1957) with the constraints on total wealth

$$\begin{aligned} p_1 + p_2 + p_3 &= 1 \\ p_1 + 4p_2 + 9p_3 &= 2 \end{aligned} \tag{5.17}$$

Solving the maximum entropy problem

$$\max\{-p_1 \ln p_1 - p_2 \ln p_2 - p_3 \ln p_3)$$

subjecting to the constraints given by Eq. (5.17) gives the following answer

$$\begin{aligned} p_1 &= 0.73 \\ p_2 &= 0.23 \\ p_3 &= 0.04 \end{aligned}$$

The total wealth for each group of investors are

$$\begin{aligned} w_1 &= 1p_1 = 0.73 \\ w_2 &= 4p_2 = 0.95 \\ w_3 &= 9p_3 = 0.33 \end{aligned}$$

Roughly speaking, small investors with 1 unit of wealth represent individual investors. Empirical evidence shows that individual investors as a group lose money from their trading activities. So we will assume small investors do not possess private information. They base their trading decisions on past price movements and other public information. Empirical evidence shows that large investors as a group make money from their trading activities. We assume the value of judgment from midsized investors with 4 unit of wealth is moderately positive and the value of judgment from large investors with 9 unit of wealth is significantly positive. The specific values of judgment by different investors will be quantified in the next paragraph.

From calculations performed in the last section, the equilibrium levels of p, d and the average proportion of risky asset held by an investor are 0.55, 0.24 and 0.4 respectively. However, from time to time, the probability of price movement of the risky asset will deviate from the equilibrium level because of various reasons. Investors with higher wealth level can detect more valuable information. Specifically, we assume large investors with 9 unit of wealth can detect all information $\{p, 1 - p\}$ with p up to 0.6 and midsized investors with 4 unit of wealth can detect all information $\{p, 1 - p\}$ with p up to 0.575.

We follow the standard literature on the assumptions of price movement of securities. The price of the risk free asset is assumed to be constant. The price movement of the risky asset is proportional to net active trading by the investors, except at the time of public release of information, when price can move without trading.

Now we consider a trading process that lasts for four time periods. At the beginning of period one, the firm underlying the risky security starts a new project. The earning from this project will become known to the public at the end of period two. This new project is expected to generate profit that corresponds to a payoff in security that is either $1 + d$ with probability 0.6 or $1 - d$ with probability 0.4 in one time period if information is publicly revealed. This expected payoff is higher than the equilibrium payoff of either $1 + d$ with probability 0.55 or $1 - d$ with probability 0.45. Large investors with 9 unit of wealth detect this information and purchase additional shares of the risky security. The proportion of wealth they invest in the risky security after the purchasing, according to formula (5.12), is

$$\frac{2p - 1}{d} = \frac{2 \times 0.6 - 1}{0.25} = 0.8$$

Since the total wealth of this group of investors is

$$w_3 = 0.33$$

The total volume of buying, which is the new holding minus the equilibrium holding at 40 %, is

$$0.33 \times (0.8 - 0.4) = 0.13$$

The purchasing by large investor increases the price of the shares and reduces the future expected returns. When the price increases to a certain level, this security will be represented by a state of future payoff that is either $1 + d$ with probability 0.575 or $1 - d$ with probability 0.425 by the end of period two. This is the end of period one and the beginning of period two. In period two, midsized investors with 4 unit of wealth detect this information and purchase shares of the risky security. The proportion of wealth they invest in the risky security, according to formula (5.12), is

$$\frac{2p-1}{d} = \frac{2 \times 0.575 - 1}{0.25} = 0.6$$

Since the total wealth of the midsized investors is

$$w_2 = 0.95$$

The total volume of their buying, which is the new holding minus the equilibrium holding at 40 %, is

$$0.95 \times (0.6 - 0.4) = 0.19$$

In period two, large investors with 9 unit of wealth will also keep 60 % of their wealth in the risky asset. As a result, they will reduce the original holding. The total volume of their selling is

$$0.33 \ \times \ (0.6 - 0.8) = -0.066$$

At the end of period two, the earning from the project becomes publicly known and the share price of the risky asset fully reflects the underlying fundamentals. Small investors with 1 unit of wealth do not possess private information. Instead, they observe that both the share price movement in the last two time periods and the earning announced at the end of period two are higher than the average. They extrapolate the past results to the future and invest accordingly. Because the share prices have moved up steadily over the last two time periods, it will be natural for small investors to base the trading decisions on the best trading decisions from two periods earlier. Specifically, in period three, on average, small investors will allocate eighty percent of their assets in the risky security. Since total wealth for the small investors is

$$w_1 = 0.73$$

The total volume of buying by the small investors, which is the new holding minus the equilibrium holding at 40 %, is

$$0.73 \ \times \ (0.8 - 0.4) = 0.29$$

Now we will consider the trading activities of large and midsized investors. Their trading decisions are based on the information they received. By default, we assume no new information in the future. In this case, the movement of stock price will return to its equilibrium condition. As a result, the holdings of the risky asset by large and midsized investors will return to the equilibrium state of forty percent. The total amount they will sell is

$$(0.33 + 0.95) \times (0.4 - 0.6) = -0.25$$

The net active trading by all investors is

$$0.29 - 0.25 = 0.04 \tag{5.18}$$

Because of the small net active trading in period three, the price movement in this period is moderate. However, trading at the beginning of period three could be dominated by small investors who mainly depend on easy to understand information, such as earning data, which is distributed widely to the general public at very narrow time frames. So trading by small investors is highly correlated (Barber et al. 2009b). Trading decisions by large and mid size investors depend more on intangible information, which is the main determinant of future returns (Daniel and Titman 2006). But intangible information is less precisely defined and trading activities generated by intangible information is less concentrated. This means that the beginning of the period three is marked by rise of asset prices while prices decline over the rest of period three. This is consistent with the empirical evidence (Hvidkjaer 2008; Barber et al. 2009a, b).

In period four, most relevant information has been acted upon and share price will finally reach equilibrium. Since share price at the end of period two has already fully reflect the fundamentals, the expected price level at the end of period four will be equal to the share price at the end of period two. Therefore, the combined net active trading of period three and four should be zero. From (4.18), the net active trading in period four should be -0.04. Since large and midsized investors already balanced their portfolio to equilibrium state in period three, the active trading is mainly generated by small investors who are reducing their holding from last period's buying. As there is little new information to generate extra trading, the total trading can be approximated by the net active trading.

We can summarize the trading activities in the four time periods into the following table:

	Period one	Period two	Period three	Period four
Net trading	0.13	0.12	0.04	−0.04
Trading volume	0.13	0.19	0.29	0.04

The average net trading of the four periods is

$$\frac{1}{4}(0.13 + 0.12 + 0.04 - 0.04) = 0.06$$

Since the net trading of the first two periods are higher than the average, share prices increase in the first two periods are higher than the average. They are the winner periods. In the last two periods, the net trading is lower than the average. Share price changes in the last two periods are lower than average. They are the loser periods. Among the winner periods, the trading volume of the first period is lower than that of the second period. Among the loser periods, the trading volume of the fourth period is lower than that of the third period. The four trading periods can be summarized as

Period one	Period two	Period three	Period four
Low volume winner	High volume winner	High volume loser	Low volume loser

This is exactly the same as the empirical pattern documented in Lee and Swaminathan (2000), which they call momentum life cycle.

Hvidkjaer (2006) examined the trading behaviors of investors of different sizes at the stages of low volume winner, high volume winner, high volume loser and low volume loser. He inferred the background of investors from the sizes of the trades. In his classification, large trades are two times or more as large as small trades. In our model, the midsized and large investors are four and nine times larger than the small investors. So it is natural to merge the midsized and large investors into one group as large investors when comparing our theoretical predictions to the empirical results documented in Hvidkjaer (2006). We will examine how trading patterns predicted from our model correspond to empirical patterns. The clearest resemblance between the predictions of our model and the empirical patterns occur in high volume loser stage. From Fig. 2 of Hvidkjaer (2006), small investors are active buyers while large investors are active sellers in this stage, which is exactly what the model has predicted. Our results are also consistent with Feng and Seasholes (2004), who showed that informed investors are selling while uninformed investors are buying after information release. In the low volume loser stage, from Fig. 3 of Hvidkjaer (2006), small investors are more active sellers than large investors. In the low volume and high volume winner stages, from Figs. 2 and 3 of Hvidkjaer (2006), large investors are more active buyers than small investors. If we interpret trading activities calculated from our model as dominant activities instead of all activities, the predictions of our model during these stages are consistent with the empirical patterns.

Alternatively, we can refine the model to make it more realistic. We had assumed the level of informedness of an investor is determined only by his wealth. To be more consistent with reality, we now assume the level of informedness of an investor is positively correlated but not determined by his wealth. Specifically, the correlation between wealth and level of informedness is represented by the following matrix.

	0.55	0.575	0.60
1	0.7	0.2	0.1
2	0.15	0.7	0.15
3	0.1	0.2	0.7

This means that among investors with one unit of wealth, 70 %, are uninformed, as 0.55 is the equilibrium state, 20 % are informed at the level of 0.575 and 10 %

are informed at the level of 0.60. The level of informedness of group two and three investors can be understood similarly.

We make a further refinement about the informedness of investors who can detect the information $\{p, 1 - p\}$ with p up to 0.6. We will assume these investors detect the information but do not interpret the information precisely. To be more specific, these investors make a judgment that p is equal to 0.575 instead of 0.6. This is very natural since most investors underestimate the significant of new information. Note that the judgment of this group of investors is still more valuable than the group of investors who estimate p to be 0.575 when it is actually 0.575. With the refined model, we can recalculate the trading activities following the same procedure as before. But this time we will measure the trading activities of large investors and small investors separately. The calculation of net trading by small and large investors at period four is determined by the proportional holdings of small and large investors at the end of period three. The results are shown in the following table.

	Period 1	Period 2	Period 3	Period 4
Small investor net trading	0.0145	0.0291	0.1600	−0.0076
Large investor net trading	0.0742	0.1455	−0.1498	−0.0026
Total net trading	0.0887	0.1746	0.0102	−0.0102
Trading volume	0.0887	0.1746	0.2735	0.0102

From the above table, the trading patterns of small and large investors calculated in each period are qualitatively similar to the empirical patterns recorded in Figs. 2 and 3 of Hvidkjaer (2006). However, we would not expect the patterns predicted from our model to be identical to empirical patterns collected in the literature. Our model presents an investment cycle initiated by a positive information signal. The empirical patterns are combinations of all kinds of cycle and non-cycle activities. For example, when the news is negative, a similar pattern exists at opposite directions. Different cycles have different amplitudes and length. In the future, we may conduct empirical investigations by filtering out different cycle and non-cycle components. This could help detect investment strategies with high level of returns.

This theory of judgment based model captured many stylized patterns of trading activities during the momentum reversal cycle. The mathematics involved are very simple and the intuition from the model is very clear. However, it is still in an early stage of development. Many refinements can be made in the future, some of which are listed as follows.

First, relation between earning momentum and price momentum can be added into the model. Empirical evidence shows strong relation between earning and price movement (Lee and Swaminathan 2000; Chordia and Shivakumar 2006). By modeling earning process over several periods of time, we can further clarify the trading mechanisms of small investors. If earning trends last longer, small investors, as well as other investors, will be more confident that the momentum will continue.

Second, informed investors can anticipate and influence the trading activities of uninformed investors. When informed investors can anticipate the trading behaviors of uninformed investors, they may base their trading decisions not only on the fundamental information, but also on the expected trading activities of uninformed investors. Informed investors, who usually have strong track records and are major stakeholders of publicly listed companies, can influence uninformed investors in certain ways to alter the trajectories of price movement to benefit themselves. These investor activities may be captured by more refined models.

5.9   Concluding Remarks

This chapter presents an updated version of the entropy theory of the human mind. Comparing with other quantitative theories in behavioral finance, which are mostly utility theories in one form or another, the entropy theory of mind is natural, simple, specific and intuitive. The entropy theory of mind is a natural extension from the entropy theory of information and statistical mechanics. The mathematical tools of the entropy theory of mind involve only simple algebraic functions such as logarithm functions and occasional use of calculus. The predictions derived from entropy theory of mind are more specific than from other theories. The links between investors' information processing and trading decisions are very intuitive under the entropy theory of mind.

Given many advantages of the entropy theory of mind over the utility theory, one may naturally ask why the concept of entropy has only very limited applications in social sciences. The question even attracted people from outside academia. William Poundstone, a writer, devoted two books to examine the historical development of the concepts of utility and entropy and their relations with human mind and investment (Poundstone 2005, 2010). Academic research, like all other social activities, is constrained by institutional structures. The concept of utility was introduced into economic research much earlier than the concept of entropy. It is always difficult to compete with a well-established theory. We can only hope, through the dedicated efforts from people both inside and outside academia, the potential of entropy theory, which has catalyzed many scientific breakthroughs in the last one and half centuries, can be further explored in the field of social sciences.

Chapter 6
The Entropy Theory of Value: A Mathematical Theory

6.1 Introduction

Value theory occupies a peculiar position in the development of economic theory. Most of the time, it is a little treated area of mainstream economists for it is generally thought to be completely resolved. But a major shift in economic thinking often begins with the emergence of new understanding about value. For example, Mill (1871) asserted that he had left nothing in the laws of value for any future economist to clear up, shortly before Jevons and Walras, in the 1870s, developed new theories of value that became the core of neoclassical economics. After neoclassical economics firmly established its dominance, research in the area of value theory became essentially dormant again in the last several decades.

There are three main theories of value: utility theory, scarcity theory and labor theory. Walras, the chief architect of the neoclassical economic theory, argued that value is a function of scarcity. He said that it is too broad to define utility as value for many things with high utility, such as oxygen, is of no economic value. It is too narrow to define labor as value, for many things take little labor have high value. For example, oil produced in Alberta takes much more labor than oil produced in Saudi Arabia. However, Alberta oil is not more expensive than Saudi oil. In conventional economic theories, additional terminologies are created, such as rent, to explain this phenomenon. However, this makes the labor theory of value less general.

Since 1870s, the dominant theory of value has been the marginal utility theory. The problem with this approach is that it completely gives up any attempt to provide a quantitative measure of value. In this chapter, we will show that the scarcity theory of value can be represented by a simple mathematical theory as an entropy theory of value. From the properties that the value of commodities should satisfy, it can be derived that the only mathematical formula to represent value, as a function of scarcity, is the entropy function. This is parallel to the idea that the only mathematical formula to represent information, as a function of probability, is the

© Springer Science+Business Media New York 2016
J. Chen, *The Unity of Science and Economics*,
DOI 10.1007/978-1-4939-3466-9_6

entropy function (Shannon 1948). The entropy theory of value is largely unchanged since its early development. This chapter is a minor update of Chapter Two of my 2005 book: They Physical Foundation of Economics: An Analytical Thermodynamic Theory.

Since all human activities represent extraction and transformation of low entropy from the environment, it is natural to relate economic value to low entropy. Indeed "there have been sporadic suggestions that all economic values can be reduced to a common denominator of low entropy" (Georgescu-Roegen 1971, p. 283). However, Georgescu-Roegen thought that linking economic value to low entropy would not be of much help to economists because "he would only be saddled with a new and wholly idle task—to explain why these coefficients differ from the corresponding price ratios" (Georgescu-Roegen 1971, p. 283). To this argument we may reflect on Shannon's entropy theory of information. The entropy theory of information does not resolve all problems related to information. But it does resolve some important problems in communication. For example, the entropy theory of information provides a measure on the minimal cost of information transmission. Very often, video data can be compressed one hundred times in transmission with little loss of quality. The entropy theory of information provides a theoretical foundation to help us transmit large amount of information at low cost, which is extremely important in today's society.

Similarly, the entropy theory of value does not resolve all problems in economic activities. But it greatly simplifies our understanding on how economic values are determined. Among other things, it offers a clear understanding how institutional structures affect economic value of commodities. Roughly speaking, economic value is the low entropy value of a commodity whose property rights are enforced by governments or other institutions. Since the costs and willingness to enforce property rights on different kinds of commodities are different, the levels of enforcement are different, which, among other factors, causes commodities of similar physical entropy level to be priced differently.

After Shannon developed the entropy theory of information, it is easy to envision an entropy theory of value as the formalization of Walras' scarcity theory of value. Historically, however, marginal utility theory of value, which was influenced by Jevons, was easier to define mathematically. Gradually, it becomes the standard economic theory (Debreu 1959). While marginal utility is easy to define mathematically, it is difficult to measure empirically. Indeed the current theory of value does not attempt to measure value empirically. This is reflected in the mathematical tools adopted in the theory: "In the area under discussion it has been essentially a change from calculus to convexity and topological properties, a transformation which has resulted notable gains in the generality and in the simplicity of the theory" (Debreu 1959, p. x). At the same time, the convexity and topological methods leave no room for a quantitative measure of value. By contrast, the entropy theory of value is established on a measurable mathematical function with clear physical meaning.

Since information is the reduction of entropy, an entropy theory of value is inevitably an information theory of value. The success of Shannon's entropy theory of information stimulated many research efforts in economics. However, the information theory of value, or the entropy theory of value, was not developed in economics. Very often, the direction of scientific research is shaped by the thinking of an authority. In an often cited passage, Arrow wrote, "the well-known Shannon measure which has been so useful in communications engineering is not in general appropriate for economic analysis because it gives no weight to the value of the information. If beforehand a large manufacturer regards it as equally likely whether the price of his product will go up or down, then learning which is true conveys no more information, in the Shannon sense, than observing the toss of a fair coin" (Arrow 1983 (1973), p. 138). The Shannon measure actually carries weight of information. For example, N symbols with identical Shannon measure carry N times more information than a single symbol (Shannon 1948). Similarly, the value of the information about the future price is higher to a large manufacturer than to a small manufacturer, other things being equal. Later in this chapter, we show that information as an economic commodity shares most of the important properties with physical commodities.

The rest of the chapter is structured as follows. In Sect. 6.2, we formally develop the mathematical theory of value as entropy. This part extends Shannon's (1948) classic work on information theory. The entropy theory of value provides a quantitative framework to understand how different factors affect the value of a commodity. The influence on value by factors such as scarcity, the number of producers and market size of a commodity can be understood naturally from the entropy formula of value. Since scarcity of resources, including human resources, is often regulated by institutional measures such as immigration laws and patent laws, the values of economic commodities are in great part a reflection of institutional structures. In Sect. 6.3, we utilize the results from information theory, statistical physics and the theory of evolution to discuss the relation between physical entropy and the economic value. We discuss how this entropy theory of value offers a unifying understanding of the objective and subjective theories of value. In Sect. 6.4, we discuss how informational and physical commodities share common properties in the light of this entropy theory of value. By resolving the conceptual difficulties that have confounded us for many years, we offer a unified understanding of physical entropy, information and economic value. In Sect. 6.5, we discuss the relation between economic value and social welfare. Section 6.6 concludes.

6.2   Main Properties

Value is a function of scarcity. Scarcity can be defined as a probability measure P in a certain probability space. It is generally agreed that the value of any product satisfies the following properties:

(a)   The value of two products should be higher than the value of each of them.
(b)   If two products are independent, that is, if the two products are not substitutes or partial substitutes of each other, then the total value of the two products will be the sum of two products.
(c)   The value of any product is non-negative.

The only mathematical functions that satisfy all of the above properties are of the form

$$V(P) = -\log_b P \tag{6.1}$$

where b is a positive constant (Applebaum 1996). In information theory, the base of the logarithm function is usually chosen to be two because there are two choices of code in information transmission, namely, 0 and 1 (Shannon 1948). In economics, the base b can be understood as the number of producers. In general, if the scarcity of a service or product, X, can be estimated by the probability measure $\{p_1, p_2, \cdots p_n\}$, the expected value of this product is the average of the value of each possibility, that is

$$V(X) = \sum_{i=1}^{n} p_i(-\log_b p_i)$$

Therefore, value, just as information, in its general form can be defined as entropy, given that they are the same mathematically. In the following we will discuss the properties of this simple analytical theory of value as scarcity.

1. **Scarcity and value**

Figure 6.1 is a graph of Formula (6.1), which shows that value is an increasing function of scarcity. That is why diamonds are worth more than water in most circumstances. In extreme abundance, i.e., when $P = 1$, $-\log P = 0$, the value of a given commodity is equal to zero, even if that commodity is very useful. For example, food is essential for survival. Most countries subsidize food production in various ways to guarantee the abundance of food, which causes its low economic value. This shows that economic value and social value can have divergent valuations.

We will examine the relation between scarcity and value for gold, silver and copper, three precious metals. The following table lists scarcity of three metals on earth, their negative log functions, their inverse functions and their actual metal prices at January 22, 2008, all normalized.

	Scarcity	Negative log of scarcity	Inverse of scarcity	Actual price
Gold	0.0000005	1	1	1
Silver	0.00001	0.79	0.05	0.018
Copper	0.007	0.34	0.00007	0.00024

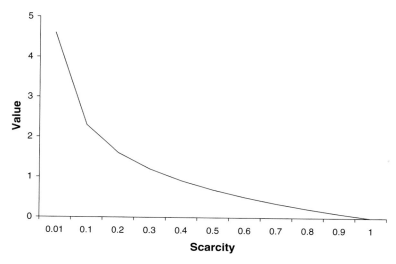

Fig. 6.1 Value and scarcity

From the above table, the negative log function seriously overvalues silver and copper prices. Inverse function more closely reflects the actual prices of these metals. These numbers show that while the negative log function qualitatively described the relation between scarcity and value, a simplistic interpretation of the function does not provide a quantitatively satisfactory answer. Are there any better alternatives that are theoretically consistent?

Gold is mined on average at much lower concentrations and takes much more energy to grind up the rocks. Likewise silver to copper. In general, a scarce commodity takes more energy and labor to mine than an abundant commodity. The scarcity theory of value is highly consistent with the energy theory of value and the labor theory of value. An advantage of the scarcity theory of value is that it can be formulated as a mathematical theory easily.

2. Value and the number of producers or consumers

From Formula (6.1), value is inversely related to the number of producers of a given product. Figure 6.2 displays the relation between value and the number of producers. When the number of producers is small, the value of a product is high. That's why the products of monopolies and oligopolies are valued highly. If the base becomes one, i.e., absolute monopoly without substitution, value approaches infinity. This happens at some religious cults where only the spiritual leaders hold the key to heaven. In these types of organizations, the leaders often enjoy tremendous amount of power over their followers. The number of providers of most economic goods depends on many factors. In the following, we give a brief discussion about the institutional structures that affect market entry and the number of suppliers for a given product.

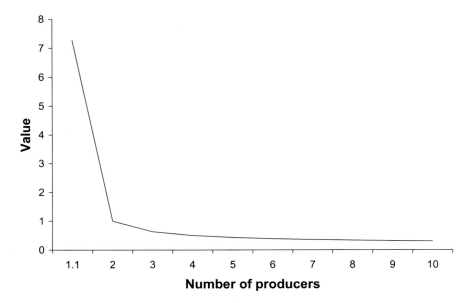

Fig. 6.2 Value and the number of producers

Anti-trust regulations aim to prevent price fixing by existing providers of a service or product. They also intend to lower barriers to potential entry. Both measures, by increasing the number of choices, reduce the value of products, and hence the cost to consumers. For this reason, the value of a product will in general be lower in a more competitive market. Patent rights and commercial secrets legislation, on the other hand, grant monopoly power and discourage the diffusion of knowledge. Patent rights and monopoly power allow the holders to maintain high product prices. The IT industry has less strict patent protection than biotech industry. As a result, the IT industry develops much faster than biotech industry. In general, industries with more patent protection develop slower than industries with less patent protection. Technology often progress very fast during war time, when patent laws are often ignored.

The quota system in trade policy forces the transfer of production technology from the dominant producer to other countries. Ultimately, the diffusion of technology and the increase of the number of producers will reduce the value of the imported goods. This will benefit the import countries over the long term, instead of the loss suggested in standard literature.

The value of consumers is also negatively related to their numbers. When there is only one dominant customer, it can mostly dictate the terms of trade and hence would like to keep the monopoly power. Producers, on the other hand, would like to increase the number of their customers. Historically, WTI and Brent crude oil prices were very close. However, WTI traded at a deep discount to Brent in recent years as Alberta increased its oil output, most of it was sold in US. In an attempt to sell more oil at the international price, proposals were made to build or expand several oil pipelines to the coastal areas so Alberta oil can be supplied to more customers. This would decrease the value and power of the existing customers and increase the value of Alberta oil products. Canada produces about three million barrels of crude oil per day. Canadian oil is often sold ten dollars per barrel below the international price. Every year the Canadian oil industry will lose about ten billion dollars from this price differential. From the perspective of the value theory, it is very easy to understand why there is so much negative publicity and disruptions around the pipeline projects.

The relation between number of producers and value can help understand many commercial and social phenomena. Each printer manufacturer designs printers in a way that printer ink from other firms cannot operate well. Customers who buy printers from one company can use ink from only the same company. By restricting the choice from customers, producers can sell ink at higher price and obtain higher profits. Workers in a profession often form unions as a single bargaining unit, which increases their negotiation power. Professions such as physicians, often are certified by a single organization, which increases their monopoly value. Doctors' notes are famously illegible. When fewer people, especially patients, are informed, the value of the profession increases. Successful religions often worship a single god instead of multiple gods. Household machines are often designed that they can only be repaired with specialized tools. Once a mixer in our home broke down. I watched a YouTube video to figure out how to have the mixer repaired. When I opened the mixer, I found the design of the mixer had been changed. With the new design, specialized tools are needed to repair the machine. A customer has to buy a new mixer or have the mixer repaired by an expensive technician. Since the value of a product depends very much on the number of producers, the attempt to gain monopoly is often the most important business strategy and political strategy (Baran and Sweezy 1966).

It is often difficult to determine the exact number of providers of a service empirically. Air travels in vast and thinly populated countries, such as Canada, where alternative modes of transportation are often very time consuming, provide a good testing ground. On March 10, 2005, Jetsgo, a Canadian airline, declared bankruptcy. There are three major operators in the air travel industry in Canada. They are Air Canada, WestJet and Jetsgo. There are regional carriers and international airlines competing for many routes. Most of the profits of airlines come from regional routes where competition is not intense. We can assume four providers for the air travel service for typical regional routes before Jetsgo declared bankruptcy. From (6.1), the value of each airline can be represented as

$$-\log_4 P \quad \text{and} \quad -\log_3 P$$

before and after Jetsgo declare bankruptcy. The change of value is therefore

$$(-\log_3 P)/(-\log_4 P) - 1 = \log_3 4 - 1 = 0.262$$

Jetsgo declared bankruptcy at the evening of March 10, 2005, after the market close. The closing prices of stocks of WestJet and Air Canada at March 10 and 11 are 11.17, 15.6 and 32.19, 37 respectively. The price changes are

$$15.6/11.17 - 1 = 0.397 \text{ for WestJet}$$

and

$$37/32.19 - 1 = 0.149 \text{ for Air Canada}$$

respectively. The average change of price is

$$(0.397 + 0.149)/2 = 0.273$$

which is very close to the theoretical prediction of 0. 262.

Some theoretical and empirical results can be further refined. For example, this theory does not distinguish the sizes of different providers of a service. The refinement of the theory is left to the future research.

3.   Market size, product life cycle and product value

Suppose the potential market size of a product is M. The percentage of people who already have the product is P. Then the unit value of the product is

$$-\log P \tag{6.2}$$

Since the number of people who have bought the product is MP, The total value of the product is

$$MP(-\log P) \tag{6.3}$$

From (6.3), the value of a product is higher with a larger market size. Figure 6.3 is the graph of unit value and total value of a product with respect to its abundance. From Fig. 6.3, we can explore the relation between the value of a product and product life cycle. When a product is new and scarce, the unit value is high. Its total value is low. As the production increases, the total value will increase as the unit value decreases. When the production quantity is over a certain level, however, the total value of a product will start to decrease as well. Intuitively, this is easy to understand. The market values of manufacturers of mature products are generally low, although the production processes are very efficient. This observation shows that efficiency is not equivalent to value.

The above discussion shows that the implications of identifying value with the scarcity are highly consistent with our intuitive understanding of economic value. It

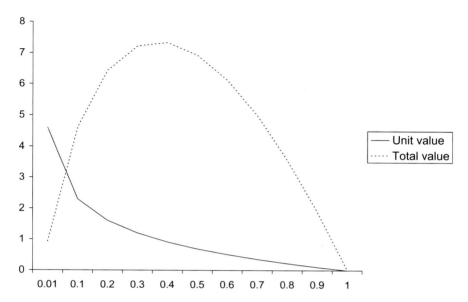

Fig. 6.3 The unit value and total value of a product with respect to scarcity

should be noted that in economic processes, a final product embodies many different kind of scarcities: labor, raw materials and equipment. A detailed analysis of the value of a particular product will be much more involved. For example, black and white television sets are less common than color television sets and yet they have less economic value. This is because the process of making color TV takes more scarce resources such as labor. The value of a final product is the combination of total scarcity.

From the value theory, product value is a function of scarcity. Tariff policy can often significantly influence output quantity and hence product value, especially when a certain commodity has one big producer and one big consumer. For example, Canada is a big producer of softwood lumber while USA is a big consumer. From value theory, the value of lumber market is represented by VP($-\ln P$), where P is the proportion of lumber that is on the market. Assume V, the total volume of the forest, is 10,000. For a consumer country, it will benefit from a trade policy that increases the production of lumber since it will reduce the value of imported lumber.

Suppose the cost structure of the lumber industry is the following. The total fixed cost in lumber production in country C is 100. The variable cost is 55 % of product value. So the total value of the lumber products is VP($-\ln P$) and the total cost of production is $100 + 0.55 * V * P * (-\ln P)$. Suppose every year, 1 % of the all lumber is harvested. The profit on lumber production is equal to revenue minus total cost

$$- VP \ln P - (100 + 0.55 * (-VP \ln P))$$
$$= -10000 * 0.01 * \ln(0.01)$$
$$- (100 + 0.55 * (-10000 * 0.01 * \ln(0.01)))$$
$$= 107$$

In 2001, USA imposes a 27 % import duty on lumber from Canada. If the volume of production remains at the same level, the profit for lumber production would be

$$- VP \ln P * (1 - 0.27) - (100 + 0.55 * (-VP \ln P))$$
$$= -10000 * 0.01 * \ln(0.01) * (1 - 0.27)$$
$$- (100 + 0.55 * (-10000 * 0.01 * \ln(0.01)))$$
$$= -17$$

which means that the lumber industry will lose money. Production of lumber has to be increased to avoid loss. If the production level is increased to $P = 1.5$ %, the profit for the lumber industry will becomes

$$- VP \ln P * (1 - 0.27) - (100 + 0.55 * (-VP \ln P))$$
$$= -10000 * 0.015 * \ln(0.015) * (1 - 0.27)$$
$$- (100 + 0.55 * (-10000 * 0.015 * \ln(0.015)))$$
$$= 13$$

As the production is increased from 1.0 % of the total reserve to 1.5 %, the unit value of lumber is decreased from $-\ln(0.01) = 4.6$ to $-\ln(0.015) = 4.2$. So USA collects 27 % tariff on lumber import and enjoy lower price on lumber. Table 6.1 is a summary statistics of softwood lumber futures price, annual production from Canada, revenues and profits from Canfor, Canada's largest softwood producer, in 2000 and 2002, one year before and after USA imposed 27 % tariff on softwood lumber import from Canada.

The empirical data confirm the theoretical predictions that after the tariff, production increased, prices dropped, and corporate profits from lumber producers tumbled. This shows that tariff is an effective way to shift wealth from producing

Table 6.1 Summary statistics of softwood lumber futures price, annual production from Canada, revenues and profits from Canfor

	2000	2002
Softwood lumber futures price (January closing)	346.6	268.7
Production (thousands of cubic meters)	68,557	71,989
Canfor revenue (millions of dollars)	2265.9	2112.3
Canfor profit (millions of dollars)	125.6	11.5

Sources of data CME, indexmundi, Canfor annual reports

countries to consuming countries and contradict the standard theory that tariff hurt import countries with higher prices for consumers.

The scarcity of a commodity is influenced by the market size. For Canadian lumber, the market size is very much determined by the US housing market. The market size is also greatly affected by the transportation cost. For example, petroleum is relatively light compared with coal for the same amount of energy. Therefore, petroleum is a global commodity while coal is much less so. Lumber is six times heavier than coal as a fuel. Hence the market size of wood as a fuel is highly localized. But the market size of wood as lumber is much larger. Still, the increasing cost of oil decreases the size of the lumber market.

6.3 Economic Value, Physical Entropy and Subjective Utility

The discussion about the relation between information and physical entropy began with the paradox of Maxwell's demon (Maxwell 1871). In 1870s, Boltzmann defined the mathematical function of entropy, which Shannon (1948) identified as information many years later. From the discussion in last chapter, information is the reduction of entropy, not only in a mathematical sense, as in Shannon's theory, but also in a physical sense. How then can economic value as entropy be linked to physical entropy?

From the last chapter, both natural selection and sexual selection indicate that human beings favor low entropy sources. This observation offers a connection between the entropy theory of value and the subjective utility theory of value. "Mind is an organ of computation engineered by natural selection" (Pinker 1997, p. 429). It calculates the entropy level and sends out signals of pleasure for accumulating and displaying low entropy and signals of pain for dissipation of low entropy. Jevons "attempted to treat economy as a calculus of pleasure and pain" (Jevons 1871, p. vi). Pleasure is generally associated with the accumulation or display of low entropy level, such as the accumulation of wealth, and conspicuous consumption. Pain is associated with dissipation of low entropy, such as work and the loss of money. So value in subjective utility theory, as a measure for pleasure and pain, is intrinsically linked to the level of entropy.

Mainstrem economic theory states that the value of a commodity is determined in exchange and is a function of supply and demand. From the theory of natural and sexual selection, the demand of an economic commodity is driven by its level of entropy. The supply of an economic commodity is constrained by its scarcity, with entropy as its unique measure. Therefore the level of entropy offers a natural measure of economic value.

It is easy to understand the objective theory of value from the entropy theory of value. Since the entropy level of a system increases spontaneously, the reduction of entropy in a system represents the effort that has been made. Entropy level may be the closest to an invariant measure of value of labor and other commodities. While economic values of commodities are highly correlated with the level of physical entropy, they are not identical for several reasons. In the following, we will discuss two: One from the perspective of information theory and another from the institutional structures that regulate scarcity and number of producers.

First, the entropy level we perceive of a commodity is different from its objective entropy level. From information theory, the amount of information one can receive, R, is equal to the amount of information sent minus the average rate of conditional entropy.

$$R = H(x) - H_y(x) \tag{6.4}$$

The conditional entropy $H_y(x)$ is called the equivocation, which measures the average ambiguity of the received signal (Shannon 1948). From our discussion in the last chapter, equivocation arises because receivers don't have the complete background knowledge of signals. For example, gold, a scarce commodity, is highly valuable. Another commodity could be as scarce as gold, but unlike shiny and stable gold, it could be very difficult to identify. Most people will not invest much effort to gain knowledge needed to identify this commodity because the cost outweighs the potential benefit. Thus, it registers less attention and is valued less by human beings.

Second, scarcity of a commodity is regulated by the institutional structures that enforce property rights. For example, the value of an invention is influenced by how long and how broad patent protection is granted. The value of a patent is higher in a system where patents are valid for twenty years than one for ten years. If patent protection is defined broader, the market size is larger and the value of an invention is higher. Economic value, as a function of scarcity, is to a great extent regulated by institutional structures. Among all the institutional measures that regulate scarcity, the most important regulation is the immigration laws that regulate the scarcity of labor forces, which makes persistent wage differential across regions possible.

6.4   The Entropy Theory of Value and Information

Because of the equivalence of entropy and information, an entropy theory of value is inevitably an information theory of value. Information is often regarded as a rather unusual commodity. In this section, we will show that informational and physical commodities share most of the fundamental properties from the perspective of entropy theory. Since Arrow (1999) offers an authoritative description about the special characteristics of information as an economic commodity, our discussion is based on his writing.

> The algebra of information is different from that of ordinary goods. ... Repeating a given piece of information adds nothing. On the other hand, the same piece of information can be used over and over again, by the same or different producer(s). (Arrow 1999, p. 21)

From Formula (6.4), the amount of information received is the information of source minus equivocation. Repeating a signal of information helps reduce equivocation. Repetition is the most important method in learning. Reciting poems is one of the most effective ways to study a language or literature. Important genes often have several hundred copies in genetic codes to satisfy heavy work demand (Klug and Cummings 2003). A song will survive only if some people repeat the lyrics. A theory will survive only if some people repeat its results. Same commercials are repeated many times on TV. A more detailed analysis of commercials by a company, say Coca Cola, will illustrate the concept more clearly. Most commercials of Coca Cola spread the same information: Drink Coca Cola. The purpose of the commercials is to reduce the equivocation in information transmission between the sender, Coca Cola company and the receivers, the potential consumers. Usually the same commercial will be repeated many times and different commercials are designed to relate the viewers to Coca Cola in different ways. However, the efforts of Coca Cola will not automatically reduce the equivocation between the sender and the receivers. Other soft drink companies and other matters in life compete for attention. As a result, the equivocation between Coca Cola and the general public may increase, despite the efforts from Coca Cola. From the thermodynamic theory that all low entropy sources have a tendency to diffuse, repeating the same piece of information is essential to keep it valuable. The essence of a living organism is to repeat and spread the information encoded in its genes.

It is often thought that the use of information does not involve rivalry, since "the same piece of information can be used over and over again, by the same or different producer(s)". This property is not confined to information. The same hammer "can be used over and over again, by the same or different producer(s)". However, the value of the same information will be different for different users or at different time. For example, if an unexpected surge of corporate profit is known by very few people, i.e., when P is very small and $-\log P$ is very high, this information would be highly valuable. Huge profit could be made by trading the underlying stocks. But when it is known to many people, the value of such information is very low. In general, when some knowledge is mastered by many people, its market value is very low.

> The peculiar algebra of information has another important implication for the functioning of the economic system. Information, once obtained, can be used by others, even the original owner has not lost it. Once created, information is not scarce in the economic sense. This fact makes it difficult to make information into property. It is usually much cheaper to reproduce information than to produce it in the first place. In the crudest form, we find piracy of technical information, as in the reproduction of books in violation of copyright. Two social innovations, patents and copyrights, are designed to create artificial scarcities where none exists naturally, although the duration of the property is limited. The scarcities are needed to create incentives for undertaking the production of information in the first place. (Arrow 1999, p. 21)

Information is a type of low entropy source. Utilization of low entropy source from others is a universal phenomenon of living systems.

Once again animals discover the trick first. ... butterflies, did not evolve their colors to impress the females. Some species evolved to be poisonous or distasteful, and warned their predators with gaudy colors. Other poisonous kinds copied the colors, taking advantage of the fear already sown. But then some nonpoisonous butterflies copied the colors, too, enjoying the protection while avoiding the expense of making themselves distasteful. When the mimics become too plentiful, the colors no longer conveyed information and no longer deterred the predators. The distasteful butterflies evolved new colors, which were then mimicked by the palatable ones, and so on. (Pinker 1997, p. 501)

So the perceived uniqueness of copying information products in human societies is actually quite universal within living systems. Once we look at the living world from the entropy perspective, it can hardly be otherwise. In human societies, the attempt to copy and reproduce valuable assets, whether informational or physical assets, is also universal.

The fashion industry offers an example that illustrates the dynamics of innovation and copying clearly. When a new fashion style is created, it is scarce and hence valuable. This valuable information will then be copied by others. As more people copy the style, P increases, $-\log P$ decreases and the value of the fashion decreases. To satisfy the demands for high value fashions, new fashion styles "are designed to create artificial scarcities where none exists naturally".

Protection of an organism's source of low entropy to prevent access by others is also a universal phenomenon of living systems. Animals develop immune systems to protect their low entropy source from being accessed by microbes. Plants make themselves poisonous to prevent their low entropy from being accessed by animals. When space is a limiting factor in survival or reproduction, animals defend their territory vigorously (Colinvaux 1978). Whether to enforce the property rights depends on the cost of enforcement and the value of the low entropy source. When information products become an important class of assets, the property rights of physical assets are naturally extended to informational assets.

6.5   Economic Value and Social Welfare

From the above discussion, it is clear that economic value and social welfare are two distinct concepts. Economic activities provide low entropy sources for the survival and comfort of human beings. From the second law of thermodynamics, the reduction of entropy locally is always accompanied by the increase of greater amounts of high entropic waste globally. So "externalities" are not a form of "market failure" but a direct consequence of fundamental physical laws. Since a low entropy product is more concentrated while high entropy waste is more diffuse, the economic value of a product is easier to measure than the harmful effects of the wastes. Usually, a product is developed to satisfy certain market demand. Its value is easily appreciated by the customers, who are willing to pay for the product. The harmful effects of the wastes, being more diffuse, affect more people but usually at very low level. These effects often take very long time to get noticed. When the human population density and consumption level is low, most of the high entropy wastes that humans generate are absorbed by microbes and other natural forces with little human effort. This vital recycling business is accorded no economic value. As the population density and the level of consumption increases, however, direct human intervention is needed to move the high entropy waste away from where people reside. The economic value of the waste management business is a function of the level of effort invested by the public in the recycling of wastes. This value is not equivalent to environmental quality of human habitats. Since high entropy waste is more diffusive, the market of recycling businesses is generally created by legal and regulatory methods to prevent the degeneration of the environment.

While economic wealth is not equivalent to social welfare, economic value, as a reflection of human efforts, is generally geared toward human welfare over the short term. This is why economic prosperity is often consistent with the improvement of social welfare in a particular point of time. However, wealth, as low entropy of human society, is ultimately supported by low entropy from nature. In the last several hundreds of years, worldwide consumption of energy has been increasing steadily with the economic progress (Smil 2003). Since our current civilization is based on fossil fuel, the eventual depletion of fossil fuel will shake the foundation of today's lifestyle (Lambert et al. 2014).

In general, wealth represents the total dependence of each other in a society. The increase of one's wealth means the increase of the dependence of others on him or her and hence the increase of his or her power. While it is natural for an individual, a company or a nation to pursue strategies that maximize wealth, such strategies ultimately will undermine long term sustainability of ecological and social systems.

6.6   Concluding Remarks

Theories built on a sound physical foundation often provide simple and intuitive results on practical problems. Shortly after Shannon's work of 1948 that identified information as entropy, Weaver commented, "Thus when one meets the concept of entropy in communication theory, he has a right to be rather excited—a right to suspect that one has hold of something that may turn out to be basic and important" (Shannon and Weaver 1949, p. 13). The phenomenal growth of information technology in the last half century has validated his foresight. This entropy theory of value, which establishes an explicit link between economic value and physical entropy, offers a simple analytical theory that greatly clarifies our understanding of economic activities.

Epilogue
Pioneer Species and Climax Species

The society tends to define scientific research as individual activities. Credits of theories go to individual authors. Patents are awarded to specific individuals and companies. But most scientific progresses and regresses are determined by the social environment. A look at interactions of different species in nature will help us understand academic and social dynamics.

Biologists often classify species into pioneer species and climax species. Pioneer species, like alders and clovers, love the sunshine, and both fix nitrogen, an energy intensive process. As they absorb solar energy to generate nutrients for themselves, they enrich the soil around them. In newly disturbed lands with poor soil content, pioneer species flourish. As pioneer species toil away, the soil they inhabit becomes fertile and rich. Over time, climax species gradually move in. Climax species, like lawn grasses and spruces, require nutritious soil. But they require less solar energy to generate nutrients themselves. So they can tolerate shade. As a forest becomes dense, new plants have to stay in the shade for a long time, waiting for old trees to retire. Pioneer species, which need an abundance of sunshine to generate nutrients, could not wait long. Gradually, climax species take over the landscape. Pioneer species are forced to look for new places to survive. This is why pioneer species are also called refugee species by biologists. On islands in northern region, poplars, a type of pioneer species, grow on the edge. Spruces grow at the core of the islands. The locations of poplars and spruces form a distinct image of periphery and core relation between pioneer and climax species.

We love climax species and often detest pioneer species. Fertilizers, which often contain herbicides to kill pioneer species, are applied lavishly on green lawns. Forestry industries encourage the growth of climax species, like spruces, and suppress the growth of pioneer species, like alders. But why?

Firstly, an abundance of climax species clearly signals the richness of the land and of the landowner. While pioneer species improve soil quality, they also indicate that the land needs to be improved, which is a sore embarrassment to proud land owners.

Secondly, pioneer species do not make high quality products. Because their priority is to generate nutrients, they dedicate less energy to their immune systems.

© Springer Science+Business Media New York 2016
J. Chen, *The Unity of Science and Economics*,
DOI 10.1007/978-1-4939-3466-9

This vulnerability leads to frequent invasion by other organisms, which lead to unsightly scars on their bodies. Hence, alders are deemed low quality timbers.

In their openness and inability to defend themselves, pioneers are more likely to accept different genes and ideas from distant species and camps. This allows pioneers to adapt, innovate and initiate major changes.

In times of poverty and turmoil, pioneers are often enlisted to help. In poor areas, farmers who cannot afford chemical fertilizers rotate the growing season of crops and clover, which is a pioneer species that enrich the soil. During the Second World War, the British government found that Alan Turing, a pioneer in understanding how the mind works, can help decoding enemy messages. But after the war, the British government found Alan Turing himself needed help. Turing rejected government's help and killed himself.

As a landscape matures, climax species become dominant and biodiversity declines. The progress of a system is often marked by the dominance of climax species over pioneer species. In modern societies, many medical and public health measures, such as sterilization and antibiotics, aim at eliminating microbes, which are often pioneer species. The short term benefits of these measures are very significant. But long term harms of many of these measures are gradually being recognized. Nowadays, doctors are less eager to prescribe antibiotics to patients. Over-sterilization deprives us from encountering stimulations that aid us in developing strong immune systems. As the environment becomes sterile, our bodies and minds follow suit. The social groups that demand highly sterile natural and social environments often suffer from below-replacement fertility rates, rendering themselves biologically sterile. If academic and social institutions continue to sterilize the social environment, the aging society will progress into a dying society, a pattern that has occurred consistently in the history of human societies.

More than ten years ago, Galbraith (2000) discussed the social environment of the economist profession and the prospect of a theoretical revolution. He wrote:

Leading active members of today's economics profession ...... have joined together into a kind of politburo for correct economic thinking. As a general rule–as one might expect from a gentleman's club–this has placed them on the wrong side of every important policy issue, and not just recently but for decades. ......

And when finally they sense that some position cannot be sustained, they do not re-examine their ideas. Instead, they simply change the subject. No one loses face, in this club, for having been wrong. No one is disinvited from presenting papers at later annual meetings. And still less is anyone from the outside invited in. ......

The reduction of many of today's leading economists to footnote status is overdue. But would those economists recognize a theoretical revolution if one were to occur? One is entitled to doubt it. Being right doesn't count for much in this club.

For a theoretical revolution to occur and to flourish, it depends on the efforts of many people.

Bibliography

Ainslie, G. (1992). *Picoeconomics: The interaction of successive motivational states within the person*. Cambridge: Cambridge University Press.

Ainslie, G., & Herrnstein, R. (1981). Preference reversal and delayed reinforcement. *Animal Learning and Behavior, 9*, 476–482.

Akerlof, G. (1970). The market for 'Lemons': Quality uncertainty and the market mechanism. *Quarterly Journal of Economics, 84*, 488–500.

Aleksandrovich, K. V., & Viktorovich, G. A. (2014). Connection of subjective entropy maximum principle to the main laws of psych. *Science and Education, 2*(3), 59–65.

Anderson, P. (2000). Cues of culture: The basis of intercultural differences in nonverbal communication. In Larry Samovar & Richard Porter (Eds.), *Intercultural communication; A reader*. Belmont, CA: Wadsworth Pub. Co.

Annila, A. (2009). Economies evolve by energy dispersal. *Entropy, 11*, 606–633.

Aoki, M., & Yoshikawa, H. (2006). *Reconstructing macroeconomics: A perspective from statistical physics and combinatorial stochastic*. Cambridge: Cambridge University Press.

Applebaum, D. (1996). *Probability and information, an integrated approach*. Cambridge: Cambridge University Press.

Arnott, R., & Casscells, A. (2003). Demographics and capital market returns. *Financial Analysts Journal, 59*(2), 20–29.

Arrow, K. J. (1973). *Information and economic behavior* (No. TR-14). CAMBRIDGE MASS: HARVARD UNIV.

Arrow, K. (1999). Information and the organization of industry. In G. Chichilnisky (Ed.), *Markets, information, and uncertainty*. Cambridge: Cambridge University Press.

Atkins, P. (1991). *Atoms, electrons, and Change*. New York: Scientific American Library, A division of HPHLP.

Atkins, P. (1997). *The periodic kingdom: A journey into the land of the chemical elements*. New York: Basic Books.

Ayres, R., van den Bergh, J., Lindenberger, D., & Warr, B. (2013). The underestimated contribution of energy to economic growth. *Structural Change and Economic Dynamics, 27*, 79–88.

Baierlein, R. (1999). *Thermal physics*. Cambridge: Cambridge University Press.

Baran, P., & Sweezy, P. (1966). *Monopoly capital*. New York: Monthly Review Press.

Barber, B. M., & Odean, T. (2000). Trading is hazardous to your wealth: The common stock investment performance of individual investors. *The Journal of Finance, 55*(2), 773–806.

Barber, B., & Odean, T. (2008). All that glitters: The effect of attention and news on the buying behavior of individual and institutional investors. *The Review of Financial Studies, 21*(2), 785–818.

Barber, B. M., Odean, T., & Zhu, N. (2009). Do retail trades move markets? *Review of Financial Studies, 22*(1), 151–186.

© Springer Science+Business Media New York 2016
J. Chen, *The Unity of Science and Economics*,
DOI 10.1007/978-1-4939-3466-9

Barber, B. M., Odean, T., & Zhu, N. (2009). Systematic noise. *Journal of Financial Markets, 12*(4), 547–569.

Barkow, J., Cosmides, L., & Tooby, J. (Eds.). (1992). *The adapted mind: Evolutionary psychology and the generation of culture.* Oxford & New York: Oxford University Press.

Banerjee, S., & Kremer, I. (2010). Disagreement and learning: Dynamic patterns of trade. *The Journal of Finance, 65*(4), 1269–1302.

Beck, K., & Andres, C. (2002). *Extreme programming explained: Embrace change* (2nd ed.). Boston: Addison-Wesley.

Beerling, D. (2007). *The emerald planet: How plants changed Earth's history.* Oxford: Oxford University Press.

Bennett, C. (1988). Notes on the history of reversible computation. *IBM Journal of Research and Development, 32,* 16–23.

Bernoulli, D. (1738(1954)). Exposition of a new theory on the measurement of risk. *Econometrica, 22*(1), 23–36.

Berns, G., Laibson, D., & Loewenstein, G. (2007). Intertemporal choice—toward an integrative framework. *Trends in Cognitive Science, 11,* 482–488.

Beutelspacher, A. (1994). *Cryptology: An introduction to the art and science of enciphering.* Washington, DC: Mathematical Association of America.

Black, F., & Scholes, M. (1973). The pricing of options and corporate liabilities. *Journal of Political Economy, 81,* 637–659.

Black, W. K. (2005). *The best way to rob a bank is to own one: How corporate executives and politicians looted the S&L industry.* Austin: University of Texas Press.

Bonner, B., & Wiggin, A. (2006). *Empire of debt: The rise of an epic financial crisis.* Hoboken, NJ: Wiley.

Boone, J. V. (2005). *A brief history of cryptology.* Annapolis, MD: Naval Institute Press.

Brav, A., & Heaton, J. B. (2002). Competing theories of financial anomalies. *Review of Financial Studies, 15*(2), 575–606.

Brown, D. (2006). *Angels and demons.* London: Simon and Schuster.

Brown, J. H., Burnside, W. R., Davidson, A. D., DeLong, J. P., Dunn, W. C., Hamilton, M. J., et al. (2011). Energetic limits to economic growth. *BioScience, 61,* 19–26.

Bryant, J. (2007). A thermodynamic theory of economics. *International Journal of Exergy, 4*(3), 302–337.

Buchanan, P. J. (2011). *Suicide of a superpower: Will America survive to 2025?* Basingstoke: Macmillan.

Campbell, C., & Laherrere, J. (1998). The end of cheap oil. *Scientific American, 278,* 78–83.

Cannon, W. B. (1932). *The wisdom of the body.*

Chakroun, I., & Abdelkader, H. (2010). Information costs in financial markets: Evidence from the Tunisian stock market. *The Journal of Risk Finance, 11*(4), 401–409.

Chatterjee, A., Yarlagadda, S., & Chakrabarti, B. K. (2005). *Econophysics of wealth distributions: Econophys-Kolkata I.* Milan: Springer.

Chen, J. (2003a). Derivative securities: What they tell us? *Quantitative Finance, 3*(5), C92–C96.

Chen, J. (2003b). An entropy theory of psychology and its implication to behavioral finance. In *Financiële Studievereniging Rotterdam Forum* (Vol. 6, No. 1, pp. 26–31).

Chen, J. (2004). Generalized entropy theory of information and market patterns. *Corporate Finance Review, 9*(3), 21–32.

Chen, J. (2005). *The physical foundation of economics: An analytical thermodynamic theory.* Hackensack, NJ: World Scientific.

Chen, J. (2006a). Imperfect market or imperfect theory: A unified analytical theory of production and capital structure of firms. *Corporate Finance Review, 11*(3), 19–30.

Chen, J. (2006b). An analytical theory of project investment: A comparison with real option theory. *International Journal of Managerial Finance, 2*(4), 354–363.

Chen, J. (2007). The informational theory of investment: A comparison with behavioral theories. *ICFAI Journal of Behavioral Finance, 4*(1), 6–31.

Chen, J. (2008a). The physical foundation of the mind. *NeuroQuantology*, *6*(3), 222.

Chen, J. (2008b). Ecological economics: An analytical thermodynamic theory. In R. Chapman (Ed.), *Creating sustainability within our midst* (pp. 99–116). New York: Pace University Press.

Chen, J. (2012). The nature of discounting. *Structural Change and Economic Dynamics, 23*, 313–324.

Chen, J., & Choi, S. (2009). Internal firm structure, external market condition and competitive dynamics. *Global Business and Economics Review, 11*(1), 88–98.

Chen, J., & Galbraith, J. (2011). Institutional structures and policies in an environment of increasingly scarce and expensive resources: A fixed cost perspective. *Journal of Economic Issues, 45*(2), 301–308.

Chen, J., & Galbraith, J. (2012). Austerity and fraud under different structures of technology and resource abundance. *Cambridge Journal of Economics, 36*(1), 335–343.

Chen, J., & Galbraith, J. (2012). A common framework for evolutionary and institutional economics. *Journal of Economic Issues, 46*(2), 419–428.

Chen, H., Jegadeesh, N., & Wermers, R. (2000). The value of active mutual fund management: An examination of the stockholdings and trades of fund managers. *Journal of Financial and Quantitative Analysis, 35*(3), 343–368.

Chen, P. (2010). *Micro interaction, meso foundation, and macro vitality: Essays on complex evolutionary economics*. London: Routledge.

Chen, Q. (2013). *Real options model of toll adjustment mechanism in concession contracts of infrastructure projects,* (Doctoral dissertation, The Hong Kong Polytechnic University).

Chordia, T., & Shivakumar, L. (2006). Earnings and price momentum. *Journal of Financial Economics, 80*(3), 627–656.

Chomsky, N. (1988). *Language and problems of knowledge: The managua lecture*. Cambridge, Mass.: MIT Press.

Chua, A. (2003). *World on fire: How exporting free market democracy breeds ethnic hatred and global instability*. New York: Anchor.

Clark, W. (2008). *In defense of self: How the immune system really works*. Oxford: Oxford University Press.

Cochran, G., & Harpending, H. (2009). *The 10,000 year explosion: How civilization accelerated human evolution*. New York: Basic Books.

Cohen, R. B., Gompers, P. A., & Vuolteenaho, T. (2002). Who underreacts to cash-flow news? Evidence from trading between individuals and institutions. *Journal of Financial Economics, 66*(2), 409–462.

Colinvaux, P. (1978). *Why big fierce animals are rare: an ecologist's perspective*. Princeton: Princeton University.

Colinvaux, P. (1980). *The fate of nations*. New York: Simon and Schuster.

Common, M., & Stagl, S. (2005). *Ecological economics: An introduction*. Cambridge: Cambridge University Press.

Cope, E. (1896). *The primary factors of organic evolution*. Chicago: The Open Court Publishing Company.

Cover, T. M., & Thomas, J. A. (2012). *Elements of information theory*. New York: Wiley.

Cronqvist, H., & Thaler, R. H. (2004). Design choices in privatized social-security systems: Learning from the Swedish experience. *American Economic Review, 94*, 424–428.

Cucchiella, F., & Gastaldi, M. (2010). Enterprise network and supply chain structure: The role of fit. In *Enterprise networks and logistics for agile manufacturing* (pp. 67–98). London: Springer.

Curzon, F. L., & Ahlborn, B. (1975). Efficiency of a carnot engine at maximum power output. *American Journal of Physics, 43*, 22.

Daly, H., & Cobb, J. (1994). *For the common good: Redirecting the economy toward community, the environment, and a sustainable future*. Boston: Beacon Press.

Damodaran, A. (2001). *Corporate finance; Theory and practice* (Second edn). New York: Wiley.

Daniel, K., & Titman, S. (2006). Market reactions to tangible and intangible information. *The Journal of Finance, 61*(4), 1605–1643.

Dargay, J., & Gately, D. (2010). World oil demand's shift toward faster growing and less price-responsive products and regions. *Energy Policy, 38*, 6261–6277.

Dawkins, R. (1999). *The extended phenotype: The long reach of the gene.* Oxford, New York: Oxford University Press.

Debreu, G. (1959). *Theory of value; an axiomatic analysis of economic equilibrium.* New York: Wiley.

Deffeyes, K. (2001). *Hubbert's peak: The impending world oil shortage.* Princeton: Princeton University Press.

Derman, E. (2004). *My life as a quant: Reflections on physics and finance.* New York: Wiley.

Devitt, M., & Sterelny, K. (1999). *Language and reality : An introduction to the philosophy of language.* Cambridge, Mass.: MIT Press.

Diamond, J. (1997). *Guns, germs, and steel, the fates of human societies.* New York: W.W. Norton.

Dixit, A., & Pindyck, R. (1994). *Investment under uncertainty.* Princeton: Princeton University Press.

Dixit, A. K., & Stiglitz, J. E. (1977). Monopolistic competition and optimum product diversity. *The American Economic Review, 67*, 297–308.

Edgerton, R. (1982). *Available energy and environmental economics.* Lexington, Mass: Lexington Books.

Einstein, A. (1905 (1989)). On a heuristic point of view concerning the production and transformation of light. In *The collected papers of Albert Einstein* (Vol. 2). Princeton: Princeton University Press.

Engelberg, J. E., & Parsons, C. A. (2011). The causal impact of media in financial markets. *The Journal of Finance, 66*(1), 67–97.

Farmer, J. D., Shubik, M., & Smith, E. (2005). Is economics the next physical science? *Physics Today, 58*(9), 37–42.

Feng, F. (2014). *A dyadic autoethnography of a learner of English via Chinese*, PhD thesis.

Ferreira, P., & Dionísio, A. (2012). Electoral results: Can entropy be a measure of the population dissatisfaction? *International Journal of Business and Management, 7*(4), p2.

Feynman, R. (1948). Space-time approach to non-relativistic quantum mechanics. *Review of Modern Physics, 20*, 367–387.

Feynman, R., & Hibbs, A. (1965). *Quantum mechanics and path integrals.* New York: McGraw-Hill.

Flassbeck, H., Davidson, P., Galbraith, J. K., Koo, R., & Ghosh, J. (2013). *Economic reform now: A global manifesto to rescue our sinking economies.* London: Palgrave Macmillan.

Frank, S. A. (1998). *Foundations of social evolution.* Princeton: Princeton University Press.

Frederick, S., Loewenstein, G., & O'Donoghue, T. (2004). Time discounting and time preference: A critical review. In C. Camerer, G. Lowenstein, & M. Rabin (Eds.), *Advances in behavioral economics.* Princeton: Princeton University Press.

Friston, K. (2010). The free-energy principle: A unified brain theory? *Nature Reviews Neuroscience, 11*(2), 127–138.

Galbraith, J. (2008). *The predator state: How conservatives abandoned the free market and why liberals should too.* New York: Free Press.

Galbraith, J. (2000). How the economists got it wrong. *The American Prospect, 11*(7), 14.

Galbraith, J. (2012). *Inequality and instability: A study of the world economy just before the great crisis.* Oxford: Oxford University Press.

Galbraith, J. (2014). *The end of normal: The great crisis and the future of growth.* New York: Simon & Schuster.

Galbraith, J. K. (1958). *The affluent society.* New York: Houghton Mifflin Harcourt.

Georgescu-Roegen, N. (1971). *The entropy law and the economic process.* Cambridge, Mass: Harvard University Press.

Gibbs, J. (1873a(1906)). *Graphical methods in the thermodynamics of fluids*. Collected in Scientific papers of J. Willard Gibbs. Woodbridge, CT: Ox Bow Press.

Gibbs, J. (1873b(1906)). *A method of geometrical representation of the thermodynamic properties of substances by means of surfaces*. Collected in Scientific papers of J. Willard Gibbs. Woodbridge, CT: Ox Bow Press.

Gibbs, J. (1902(1981)). *Elementary principles in statistical mechanics*. Woodbridge, Connecticut: Ox Bow Press.

Gisolfi, C., & Mora, F. (2000). *The hot brain: Survival, temperature, and the human body*. Cambridge, MA: MIT Press.

Gleick, J. (2011). *The information: A history, a theory, a flood*. New York: Pantheon.

Golden, D. (2006). *The price of admission: How America's ruling class buys its way into elite colleges-and who gets left outside the gates*. New York: Crown.

Green, L., Myerson, J., & McFadden, E. (1997). Rate of temporal discounting decreases with amount of reward. *Memory and Cognition, 25*(5), 715–723.

Gribnikov, V., & Shevchenko, D. (2012). Influence of behavioral finance on the share market. In *Market risk and financial markets modeling* (pp. 57–61). Berlin Heidelberg: Springer.

Grossman, S. J., & Stiglitz, J. E. (1980). On the impossibility of informationally efficient markets. *The American Economic Review, 70*, 393–408.

Haig, D. (1993). Genetic conflicts in human pregnancy. *Quarterly Review of Biology*, 495–532.

Hall, C. (2004). The myth of sustainable development: Personal reflection on energy, its relation to neoclassical economics, and Stanley Jevons. *Journal of Energy Resource Technology, 126*, 85–89.

Hall, C., Cutler J. C., & Robert K. (1986). *Energy and resource quality: The ecology of the economic process*. New York: Wiley.

Hall, C., & Klitgaard, K. (2006). The need for a new, biophysical-based paradigm in economics for the second half of the age of oil. *International journal of transdisciplinary research, 1*(1), 4–22.

Hall, C. A., & Klitgaard, K. A. (2011). *Energy and the wealth of nations: Understanding the biophysical economy*. Dordrecht: Springer Science & Business Media.

Hall, C. A., Lambert, J. G., & Balogh, S. B. (2014). EROI of different fuels and the implications for society. *Energy Policy, 64*, 141–152.

Hall, E. T. (1977). *Beyond culture*. Garden City, N.Y.: Anchor Press.

Hallett, S., & Wright, J. (2011). *Life without oil: Why we must shift to a new energy future*. Amherst: Prometheus Books.

Hamilton, W. D. (1998). *Narrow roads of gene land: The collected papers of WD Hamilton*. Oxford: Oxford University Press.

Hanley, R., Tzeng, O., & Huang, H. S. (1999). Learning to read Chinese. In M. Harris & G. Hatano (Eds.), *Learning to read and write*. Cambridge: Cambridge University Press.

Harcourt, G. (1969). Some Cambridge controversies in the theory of capital. *Journal of Economic Literature, 7*(2), 369–405.

Harigaya, M. (2001). Private communication.

Hawkins, R. J., Aoki, M., & Roy Frieden, B. (2010). Asymmetric information and macroeconomic dynamics. *Physica A: Statistical Mechanics and its Applications, 389*(17), 3565–3571.

Heinberg, R. (2007). *Peak everything: Waking up to the century of declines*. Philadelphia: New Society Publishers.

Henderson, Y. (1940). Carbon dioxide. In *Cyclopedia of medicine*. Philadelphia: FA Davis.

Hofstede, G. (1980). *Culture's consequences: International differences in work-related value*. Beverly Hills, Calif.: Sage Publications.

Hvidkjaer, S. (2006). A trade-based analysis of momentum. *Review of Financial Studies, 19*(2), 457–491.

Imura, H. (2013). Environmental systems studies: A macroscope for understanding and operating spaceship earth. Japan: Springer.

Inhaber, H. (1997). *Why energy conservation fails*. Westport, Connecticut: Quorum Books.

Iorgulescu, R., & Polimeni, J. (2007). The physical foundation of economics. *Ecological Economics, 62*(1), 195–196.

Jablonka, E., & Lamb, M. (2006). *Evolution in four dimensions: Genetic, epigenetic, behavioral, and symbolic variation in the history of life*. Cambridge, MA: The MIT Press.

Jalil, A. (2014). Energy–growth conundrum in energy exporting and importing countries: Evidence from heterogeneous panel methods robust to cross-sectional dependence. *Energy Economics, 44*, 314–324.

Jammer, M. (1966). *The conceptual development of quantum mechanics* (2nd ed.). New York: McGraw-Hill.

Janszen, E. (2008). The next bubble: Priming the markets for tomorrow's big crash. *Harper's, 316*, 39–45.

Jaynes, E. (1957). Information theory and statistical mechanics. *Physical Review, 106*, 620–630.

Jaynes, E. (1988). How does the brain do plausible reasoning?. In G. J. Erickson & C. R. Smith (Eds.), Maximum-entropy and bayesian methods in science and engineering (Vol. 1, pp. 1–24). Dordrecht: Kluwer.

Jaynes, E. (2003). *Probability theory: The logic of science*. Cambridge: Cambridge University Press.

Jevons, W. (1865). *The coal question: An inquiry concerning the progress of the nation, and the probable exhaustion of our coal-mines*. London: Macmillan and Co.

Jevons, W. (1871). *The theory of political economy*. London: Macmillan and Co.

Johnson, N. (2008). The revolution will not be pasteurized. *Harpers*, 71–78.

Jones, S. (2008). *The quantum ten: A story of passion, tragedy, ambition and science*. Oxford: Oxford University Press.

Kac, M. (1951).*On some connections between probability theory and differential and integral equations*. In J. Neyman (Ed.), *Proceedings of the second Berkeley symposium on probability and statistics* (pp. 189–215). Berkeley: University of California.

Kac, M. (1985). *Enigmas of chance: An autobiography*. New York: Harper and Row.

Kellogg, E. (2001). *Evolutionary history of the grasses plant physiology, 125*, 1198–1205.

Kelly, J. (1956). A new interpretation of information rate. *Bell System Technical Journal, 35*, 917–926.

Keynes, J. (1932). *Essays in persuasion*. New York: Harcourt, Brace and Company.

Klug, W. S., & Cummings, M. R. (2003). *Concepts of genetics* (7th edn). Snustard, D.P: Pearson Education, Inc.

Kolmogorov, A. (1931). *On analytical methods in the theory of probability*. (Reprinted in Selected Works of A. N. Kolmogorov, Vol. II, Springer, 1992).

Kong, X., Liu, L., & Chen, J. (2011). modeling agile software maintenance process using analytical theory of project investment. *Procedia Engineering, 24*, 138–142.

Kragh, H. (2000). Max planck, the reluctant revolutionary. *Physics World, 13*(12), 31–35.

Krugman, P. (2009). How did economists get it so wrong? *New York Times, 2*(9), 2009.

Kuhn, T. (1996). *The structure of scientific revolutions* (3rd ed.). Chicago: University of Chicago Press.

Kullback, S. (1959). *Information theory and statistics*. New. York: Wiley.

Kunstler, J. (2005). *The long emergency: Surviving the converging catastrophes of the twenty-first centur*. New York: Atlantic Monthly Press.

La Cerra, P. (2003). The first law of psychology is the second law of thermodynamics: The energetic evolutionary model of the mind and the generation of human psychological phenomena. *Human Nature Review, 3*, 440–447.

La Cerra, P., & Bingham, R. (1998). The adaptive nature of the human neurocognitive architecture: An alternative model. *Proceedings of the National Academy of Sciences, 95*(19), 11290–11294.

Lambert, J. G., Hall, C. A., Balogh, S., Gupta, A., & Arnold, M. (2014). Energy, EROI and quality of life. *Energy Policy, 64*, 153–167.

Lane, N. (2002). *Oxygen: The molecule that made the world*. Oxford: Oxford University Press.

Lane, N. (2010). *Life ascending: The ten great inventions of evolution.* US: W. W. Norton & Company.

Latane, H. (1959). Criteria for choice among risky ventures. *The Journal of Political Economy, 67*(2), 144–155.

Latane, H., & Tuttle, D. (1967). Criteria for portfolio building. *Journal of Finance, 22*(3), 359–373.

Lee, C., & Swamminathan, B. (2000). Price momentum and trading volume. *Journal of Finance, 55*(5), 2017–2069.

Legendre, T. (2006). *The burning.* New York: Little, Brown.

Leppinen, E. (2013). *Behavioral finance theories effecting on individual Investor's' decision-making,* thesis.

Lewis, M. (2011). *The big short: Inside the doomsday machine.* New York: WW Norton & Company.

Lewis, M. (2014). *Flash boys: A wall street revolt.* New York: WW Norton & Company.

Liao, W., Heijungs, R., & Huppes, G. (2012). Thermodynamic analysis of human–environment systems: A review focused on industrial ecology. *Ecological Modelling, 228,* 76–88.

Liu, G., Xu, K., Zhang, X., & Zhang, G. (2014). Factors influencing the service lifespan of buildings: An improved hedonic model. *Habitat International, 43,* 274–282.

Liu, L., Kong, X., & Chen, J. (2015). How project duration, upfront costs and uncertainty interact and impact on software development productivity? A simulation approach. *International Journal of Agile Systems and Management, 8*(1), 39–52.

Lu, X., & Zhang, J. (1999). Reading efficiency: A comparative study of English and Chinese orthographies. *Reading Research And Instruction, 38*(4), 301–317.

MacArthur, R. H., & Wilson, E. O. (1967). *The theory of island biogeography.* Princeton: Princeton University press.

Madden, B. (2014). *Reconstructing your worldview: The four core beliefs you need to solve complex business problems.* Harvard: Harvard Business School Press.

Maestripieri, D. (2012). *Games primates play: An undercover investigation of the evolution and economics of human relationships.* New York: Basic Books.

Maor, E. (1994). *e: The story of a number.* Princeton: Princeton University press.

Maupertuis, P. (1746). *Derivation of the laws of motion and equilibrium from a metaphysical principle* (p. 267). Berlin: Mém. Ac.

Maxwell, J. (1871). *Theory of heat.* Longmans, London: Green & Co.

McEnally, R. (1986). Latane's bequest: The best of portfolio strategies. *The Journal of Portfolio Management, 12*(1), 21–30.

Megginson, W. (1997). *Corporate finance theory.* Reading, Mass: Addison-Wesley.

Mehrling, P. (2005). *Fischer black and the revolutionary idea of finance.* Hoboken, NJ: Wiley.

Meyer, T., & Mathonet, P. Y. (2011). *Beyond the J curve: Managing a portfolio of venture capital and private equity funds* (Vol. 566). New York: Wiley.

Mill, J. (1985(1871)). *Principle of political economy.* London: Penguin books.

Mimkes, J., & Aruka, Y. (2005). Carnot process of wealth distribution. In *Econophysics of wealth distributions* (pp. 70–78). Milan: Springer.

Moalem, S., & Prince, J. (2008). *Survival of the sickest: The surprising connections between disease and longevity.* New York: Harper Perennial.

Murphy, D., & Hall, C. A. (2010). Year in review—EROI or energy return on (energy) invested. *Annals of the New York Academy of Sciences, 1185*(1), 102–118.

Nelson, R., & Winter, S. (1982). *An evolutionary theory of economic change.* Cambridge: Harvard University Press.

Newell, R., & Pizer, W. (2003). Discounting the distant future: how much do uncertain rates increase valuations? *Journal of Environmental Economics and Management, 46*(1), 52–71.

Nikiforuk, A. (2012). *The energy of slaves: Oil and the new servitude.* Vancouver: Greystone Books.

Novy-Marx, R. (2012). Is momentum really momentum? *Journal of Financial Economics, 103*(3), 429–453.

Odean, T. (1999). Do Investors Trade Too Much. *American Economic Review, 89*(5), 1279–1298.

Odum, H. (1971). *Environment, power and society*. New York: Wiley.

Øksendal, B. (1998). *Stochastic differential equations: an introduction with applications* (5th ed.). Berlin, New York: Springer.

Ophuls, W. (2012). *Immoderate greatness: Why civilizations fail*. Charleston, SC: Create Space.

Outreville, J. F. (2013). The price of wine: Does the bottle size matter? *Wine Economics: Quantitative Studies and Empirical Applications, 88.*

Parker, P. (2000). *Physioeconomics: The basis for long-run economic growth*. Cambridge, Mass.: MIT Press.

Pati, S. P. (2009). Stress management and innovation: A thermodynamic view. *Journal of Human Thermodynamics, 5,* 22–32.

Patterson, S. (2010). *The quants: How a new breed of math whizzes conquered wall street and nearly destroyed it*. New York: Crown Business.

Peng, K. (2005). The measurement of decision uncertainty by Shannon's entropy. *Ming Hsin Journal, 31,* 171–181.

Pierce, J. (1980). *An introduction to information theory: Symbols, signals and noise*. New York: Dover Publications.

Pimentel, D., & Patzek, T. W. (2005). Ethanol production using corn, switchgrass, and wood; biodiesel production using soybean and sunflower. *Natural Resource Research, 14*(1), 65–76.

Pinker, S. (1994). *The language instinct*. New York: W. Morrow.

Pinker, S. (1997). *How the mind works*. New York: W. W. Norton.

Planck, M. (1900). On an improvement of wien's equation for the spectrum. *Verhandlungen der Deutschen Physik. Gesells, 2,* 202–204.

Planck, M. (1901). On the law of distribution of energy in the normal spectrum. *Annalen der Physik, 4*(553), 1.

Pogany, P. (2012). Value and utility in a historical perspective, working paper.

Poundstone, W. (2005). *Fortune's formula: The untold story of the scientific betting system that beat the casinos and wall street*. New York: Hill and Wang.

Poundstone, W. (2010). *Priceless: The myth of fair value (and how to take advantage of it)*. New York: Hill and Wang.

Prieto, P. A., & Hall, C. (2013). *Spain's photovoltaic revolution: The energy return on investment*. New York: Springer Science & Business Media.

Prigogine, I. (1980). *From being to becoming: Time and complexity in the physical sciences*. San Francisco: W. H. Freeman.

Qian, H. (2001). Relative entropy: Free energy associated with equilibrium fluctuations and non-equilibrium deviations. *Physical Review E, 63,* 042103.

Qian, H. (2009). Entropy demystified: The "thermo"-dynamics of stochastically fluctuating systems. In M. Johnson & L. Brand (Eds.), *Methods in enzymology Computer methods, part B* (Vol. 467, pp. 112–134). London: Academic Press.

Qian, H. (2010). Cellular biology in terms of stochastic nonlinear biochemical dynamics: Emergent properties, isogenetic variations and chemical system inheritability. *Journal of Statistical Physics, 141,* 990–1013.

Rando, O., & Verstrepen, K. (2007). Timescales of genetic and epigenetic inheritance. *Cell, 128*(4), 655–668.

Ricklefs, R. (2001). *The economy of nature* (5th ed.). New York: W.H. Freeman.

Romm, J. (2005). *The hype about hydrogen: Fact and fiction in the race to save the climate*. Washington: Island Press.

Romer, P. (1986). Increasing returns and long-run growth. *Journal of Political Economy, 94*(5), 1002–1037.

Rubí, J. (2008). The long arm of the second law. *Scientific American, 299*(5), 62–67.

Rubin, J. (2009). *Why your world is about to get a whole lot smaller*. New York: Random House.

Rushton, P. (1996). *Race, evolution, and behavior: A life history perspective*. New Brunswick, NJ: Transaction Publishers.

Sabolovič, M. (2011). An alternative methodological approach to value analysis of regional, municipal corporations and clusters. *ACTA UNIVERSITATIS AGRICULTURAE ET SILVICULTURAE MENDELIANAE BRUNENSIS, 35*(4), 295–300.

Samuelson, P. (1969). Portfolio selection by dynamic stochastic programming. *The Review of Economics and Statistics, 51*(3), 239–246.

Samuelson, P. (1972). *The collected scientific papers* (Vol. 3). Cambridge, Mass.: MIT Press.

Samuelson, P., & Nordhaus, W. (1998). *Economics* (16th ed.). Boston, MA: McGraw Hill.

Schapiro, M. (2010). Conning the climate: Inside the carbon-trading shell game, *Harpers*, 31–39.

Schlögl, F. (1989). *Probability and heat : Fundamentals of thermostatistics*. Braunschweig: F. Vieweg.

Schmidt-Nielson, K. (1997). *Animal physiology* (fifth ed.). Cambridge: Cambridge University Press.

Schneider, E. D., & Sagan, D. (2005). *Into the cool: Energy flow, thermodynamics, and life*. Chicago: University of Chicago Press.

Schrodinger, E. (1928). *Collected papers on wave mechanics*. London: Blackie and Son Limited.

Schrodinger, E. (1944). *What is life?* Cambridge: Cambridge University Press.

Shannon, C. (1948). A mathematical theory of communication. *The Bell System Technical Journal, 27*(379–423), 623–656.

Shannon, C. (1956). The bandwagon (Editorial). *IEEE Transactions Information Theory, 2*, 3.

Shannon, C., & Weaver, W. (1949). *The mathematical theory of communication*. Urbana: The University of Illinois Press.

Shubin, N. (2008). *Your inner fish*. New York: Pantheon Books.

Sinn, H. (2003). Weber's law and the biological evolution of risk preferences: The selective dominance of the logarithmic utility function. *The Geneva Papers on Risk and Insurance Theory, 28*, 87–100.

Smil, V. (2003). *Energy at the crossroads: Global perspectives and uncertainties*. Cambridge, MA: The MIT Press.

Sobottka, S. (2005). A course in consciousness. http://faculty.virginia.edu/consciousness/

Sornette, D. (2009). *Why stock markets crash: Critical events in complex financial systems*. Princeton: Princeton University Press.

Stearns, S. (1992). *The evolution of life histories*. Oxford: Oxford University Press.

Stevens, S. (1961). To Honor Fechner and Repeal His Law: A power function, not a log function, describes the operating characteristic of a sensory system. *Science, 133*(3446), 80–86.

Stigler, S. (1986). *The history of statistics: The measurement of uncertainty before 1900*. Cambridge: The Belknap Press.

Stiglitz, J. (2002). *Globalization and its discontents*. New York: W.W. Norton & Company.

Stiglitz, J. (2010). Needed: A new economic paradigm, *Financial Times*, August 19.

Stiglitz, J. (2012). *The price of inequality*. UK: Penguin.

Tainter, J. (1988). *The collapse of complex societies*. New York: Cambridge University Press.

Ternyik, S. (2012). The monetary quantum, *working paper*.

Thaler, R. (1981). Some empirical evidence on dynamic inconsistency. *Economics Letters, 8*(3), 201–207.

Thims, L. (2007). *Human chemistry*. North Carolina: Lulu.

Thims, L. http://www.humanthermodynamics.com/

Thorp, E., & Kassouf, S. (1967). *Beat the market: A scientific stock market system*. New York: Random House.

Thorp, E. (1997). *The Kelly criterion in blackjack, sports betting and the stock market*. Presented at: The 10th International Conference on Gambling and Risk Taking Montreal.

Tooby, J., & Cosmides, L. (1992). The psychological foundations of culture. In J. Barkow, L. Cosmides, & J. Tooby (Eds.), *The adapted mind: Evolutionary psychology and the generation of culture*. New York: Oxford University Press.

Tooby, J., Cosmides, L., & Barrett, C. (2003). The second law of thermodynamics is the first law of psychology" Evolutionary development and the theory of tandem, coordinated inheritances. *Psychological Bulletin, 129*(6), 858–865.

Treynor, J. (1996). Remembering fischer black. *Journal of Portfolio Management, 23*, 92–95.

Trivers, R. (1985). *Social evolution.* Menlo Park, CA: Benjamin/Cummings.

Trivers, R. (2011). *The folly of fools: The logic of deceit and self-deception in human life.* New York: Basic Books.

Tverberg, G. http://ourfiniteworld.com/

Tversky, A., & Kahneman, D. (1974). Judgment under uncertainty: Heuristics and biases. *Science, 185*(4157), 1124–1131.

Unz, R. (2012). The myth of American meritocracy. *The American Conservative*, 14–51.

Vasantrao, K. (2011). Understanding need of flexible software development approach using "The Economic Model". In *2011 3rd International Conference on Electronics Computer Technology (ICECT)* (Vol. 5, pp. 240–243). IEEE.

Veblen, T. (1899). *The theory of the leisure class.* New York: MacMillan.

Vogel, J. H. (1989). Entrepreneurship, evolution, and the entropy law. *Journal of Behavioral Economics, 18*(3), 185–204.

Wachowicz, M., & Owens, J. (2013). The role of knowledge spaces in geographically-oriented history. In *History and GIS: Epistemologies, considerations and reflections*, edited by Alexander von Lünen and Charles Travis, Springer.

Wackernagel, M., & Rees, W. (1995). *Our ecological footprint: Reducing human impact on the earth.* New Haven, CT: New Society Publishers.

Wang, F., Cai, Y., & Gu, B. (2013). Population, policy, and politics: How will history judge China's one-child policy? *Population and development review, 38*(s1), 115–129.

Weitzman, M. (2001). Gamma discounting. *American Economic Review, 91*(1), 260–271.

Wermers, R. (2000). Mutual fund performance: An empirical decomposition into stock-picking talent, style, transaction costs, and expenses. *Journal of Finance, 55*(4), 1655–1703.

Whitfield, J. (2006). *In the beat of a heart: Life, energy, and the unity of nature.* Washington, D.C.: Joseph Henry Press.

Whorf, B. L. (1956). *Language, thought, and reality; Selected writings.* Cambridge, Mass.: Technology Press of Massachusetts Institute of Technology.

Willey, J. M., Sherwood, L. M., & Woolverton, C. J. (2011). *Prescott's Microbiology.* New York: McGraw-Hill.

Williams, G. (1966). *Adaptation and natural selection: A critique of some current evolutionary thought.* Princeton : Princeton University Press.

Wilson, D. (2007). *Evolution for everyone: How Darwin's theory can change the way we think about our lives.* New York: Random House LLC.

Wrangham, R. (2009). *Catching fire: How cooking made us human.* New York: Basic Books.

Wright, G., & Czelusta, J. (2007). Resource-based growth past and present. In D. Lederman & W. F. Maloney (Eds.), *Natural resources: Neither curse nor destiny* (pp. 183–211). Washington, DC: World Bank/Stanford University Press.

Yevdokimov, Y., & Molchanov, M. A. (2013). State regulation or state capitalism?: A systems approach to crisis prevention and management. *International Journal of Management Concepts and Philosophy, 7*(1), 1–12.

Yu, L. (2012). An empirical study of software market share: Diversity and symbiotic relations. *First Monday, 17*(8), 6 August.

Zhang, Y., Deng, J., Majumdar, S., & Zheng, B. (2009). Globalization of lifestyle: Golfing in China. In L. Hellmuth & L. Meier (Eds.), *The new middle classes, globalizing lifestyles, consumerism and environmental concern.* New York, NY: Springer.

Zipf, G. (1949). *Human behavior and the principle of least effort: An introduction to human ecology.* Cambridge/Mass: Addison-Wesley Press.